# Survival
# Fire Water

*Nature's Power*
*Simple Techniques*

by Dennis Kingston

**CLADD**
PUBLISHING

Cladd Publishing Inc.
USA

This publication is designed to provide accurate information regarding the
subject matter covered. It is sold with the understanding that neither the
author nor the publisher is providing medical, legal or other professional
advice or services. Always seek advice from a competent professional
before using any of the information in this book. The author and the
publisher specifically disclaim any liability that is incurred from the use or
application of the contents of this book.

Survival Fire Water: Nature's Power, Simple Techniques

ISBN 978-1-946881-65-6 (e-book)
ISBN 978-1-946881-64-9 (paperback)

# Contents

# Will You Live or Die?

Every person should know how to start a fire and find or make clean water. This knowledge isn't just handy to have in the event we experience a nuclear war. Being prepared gives you the ability to survive no matter what life throws your way.

Many people have died from dehydration and starvation after getting lost while hunting, mountain climbing, rock climbing, mountain biking, camping, snow skiing, recreational vehicle, car wreck, plane crash, boating incident or a physical injury that left them stranded. And of course, is the occasional earth quake, landslide, avalanche, tsunami, hurricane, tornado, plagues, nuclear plant meltdown and war.

## WE ONLY NEED 3 THINGS TO SURVIVE:
1. Water
2. Fire
3. Shelter

If you understand how to facilitate a fire, clean water and shelter, then you will survive even in the most dreadful circumstances. You will have the power and knowledge to provide your loved ones with life, when death is staring them in the eye.

By mastering the creation of fire, you will have the capability of cleaning your water, cooking food and staying warm. The rest of survival will come much easier, because you will have proper hydration and nutrients to do what is necessary thereafter.

## LET'S GET STARTED

It is critical for your survival to understand that each situation that requires you to put to use these new skills will be different. You could be traveling home from work, and an emergency warning interrupts your radio station declaring nuclear war. The power-grid could be attacked and power-outages span across entire States or Countries. There could be a catastrophic storm that wipes everything out and leaves you and your family injured and stranded unable to be helped.

These situations could seem far-fetched but let me remind you that not only have they already happened, but they are currently happening to people all around the world. The better you are prepared, the more likely you will survive in dangerous situations of any kind. Since there are so many events that could force you into survival mode, we will be teaching you how to master finding and creating clean water and starting a fire from common items. These are the essential components to keeping yourself and loved ones alive.

Having the bravery to do what is necessary, with what is available, is all that you will have when the time comes.

This book will teach you how to use odd, unusual and commonly overlooked items from homes, businesses and off the streets to create clean water, and fire. With these techniques, you will be armed with the knowledge to survive any situation.

# Finding & Making Clean Water

# We Need Water

WATER NEEDS CALCULATOR:

BODY WEIGHT lb ✕ 15 = ml

94% LYMPH

86% LIVER

83% BLOOD

83% KIDNEYS

83% JOINTS

80% LUNGS

79% HEART

75% MUSCLE

75% BRAIN

64% SKIN

22% BONES

70% WATER

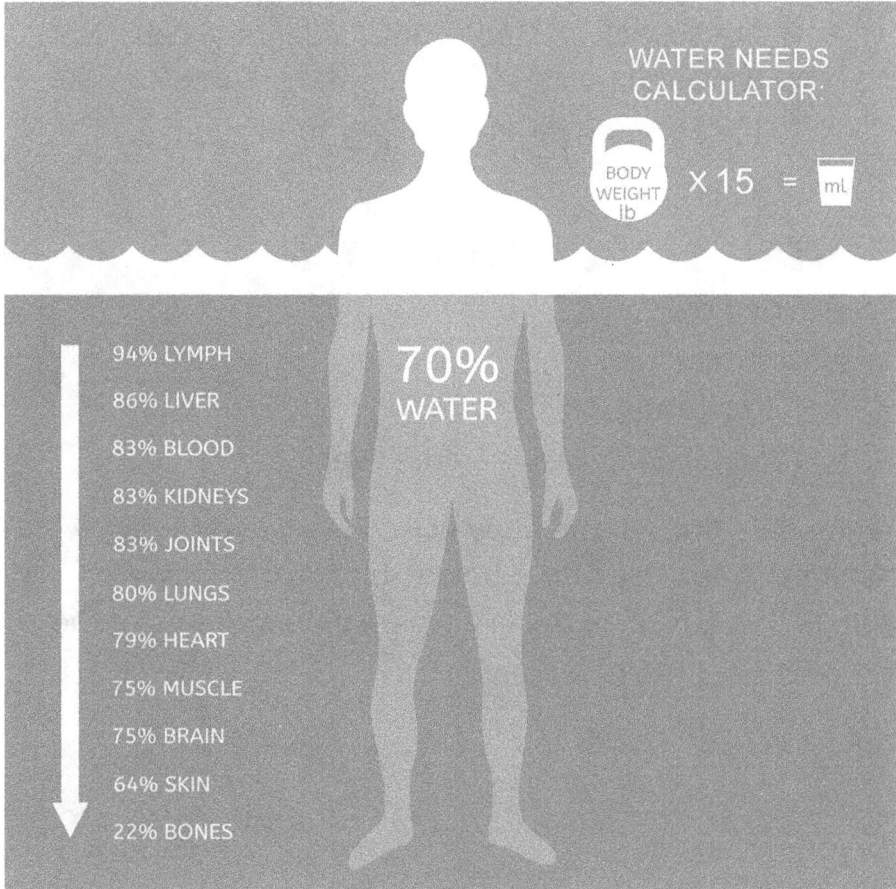

You can survive for weeks without food, but only a few days without water. Water is clearly the most important nutrient and the most abundant substance in the human body. Water equates to about three quarters of the human mass and is a major component in every cell. Unlike vitamin deficiencies which can take a long time to manifest into an illness, a water deficiency can kill in days or even hours.

# DEHYDRATION KILLS

➢ Being 2% dehydrated seriously degrades physical and mental function.
➢ Being 15% dehydrated is likely to be deadly.

# WATER ALSO AIDS OUR BODY IN:

- Bowels
- Urination
- Perspiration
- Processing of toxins by the liver

When dehydrated, the body will conserve water by minimizing the use of the bowel, bladder, and sweat glands. Thus, forcing the liver to assume as much of the workload as possible.

# Items Needed for Water Survival

### STAINLESS STEEL WATER CONTAINER
Make sure to carry the kind with a screw-on lid. It should be made for hiking. You will be able to carry water and use it to boil for purification.

### PLASTIC BAGS OR PONCHO
You will need many plastic bags that are free of tears, or a plastic poncho could be a multipurpose substitute.

### LARGE NATURAL FABRIC SQUARE
You will use these fabric squares for absorbing water to drink, straining, and larger filtration systems. You could also use clothing if necessary.

### ROPE
Rope will be valuable for building larger water filtration systems, along with securing small water trapping devices.

### TARP
Tarp is important for creating larger water filtration systems.

### FIRE STARTER
It is always vital to have heat and fire. To purify the most contaminated sources of water in large quantities, boiling will almost always be necessary.

# Finding Water Sources in Nature

## TRAVEL PARALLEL TO A MOUNTAIN

If you're a survivor and find yourself in a mountain range, locating water is going to come down to traveling parallel to a mountainside. Eventually you will cross a stream. You may even come across a pond or a lake where you can fish and find other wild edibles.

## LAKES, PONDS, RIVERS

Lakes and ponds are stagnant, which increases the levels of bacteria. While large rivers are normally full of pollution. Be careful after any flooding, or if the river flows from or through a chemical plant, under a road or around any construction.

## SNOW AND ICE

Snow and ice are a readily available source of clean water in the winter.

- **Do not eat snow or ice directly = Lowers body temperature**

In a perfect world, you should purify snow water once melted, but as long as it's not yellow or brown, consumption would be better than dehydration.

## MUD

There tends to be relatively clean water underground around small bodies of water.

## Sea Water & Urine

Never drink urine or sea water directly. However, you can boil them and collect the steam with a plastic sheet or bag.

## Branches & Foliage

Branches and Foliage offer us an excellent way of consuming water without the need for fire and other more complicated collection devices. Never do this with a poisonous plant.

## Downhill

If you cannot find any water, then start walking downhill. Look for dark patches in the landscape and any group of vegetation that stands out in lower areas.

## Trees

Cut a tree with a knife or sharp rock. Within a few minutes you can make a hole in the tree and begin pulling small amounts of water.

## Rain Water

This is the source of all clean water and is considered the top of the water chain. If it rains and you need water, then start collecting in any way possible.

# FOG

Fog is considered a cloud that touches or contacts the ground. Fog provides us with water droplets suspended in the air. The thickness of water particles is dense enough for us to trap directly out of the air. Because fog is evaporated water, its right from Mother Nature and ready to drink.

# CREVICES

In mountains ranges, water can accumulate from rain fall. Look for crevices between boulders, and especially look for crevices that lie in shadows when the sun is high.

# WATER WITCHING DOWSING

Dowsing is when a receptive person uses one or two twigs to sense or "feel" underground water. The sticks are held in each hand side-by-side loosely. After slowly walking around an area, the twigs will begin to pull when a water source is close by.

# GEOLOGICAL SURVEY MAP

Order a geological survey map of underground water areas for your location. They are normally used for homeowners wanting to have a private well drilled, but they can be a lifesaver to have on hand in case of an emergency. This is a map of actual "water areas" underground, along with the production quantities.

# Emergency Water Supplies in Your Home

After a disaster such as an earthquake, tsunami, major fire, or war; some or all supplies and services can be cut off for an undisclosed length of time.

This can also mean that water contamination becomes a real possibility. It is good to know how to locate and purify water in these emergencies.

## HIDDEN DRINKING WATER IN OUR HOMES

- **Hot Water Tanks:** Turn off the source of heat and when the tank has cooled you simply use the tap at the bottom of the tank to get your water.
- **Water Pipes:** Open a tap at the highest point possible for air to enter, and then open a downstairs (or lower) tap to get the water out.
- **Freezer Ice Cubes:** The freezer usually contains some sort of ice cube tray. Melt the cubes and drink the water.
- **Toilet Tank:** As long as you don't use chemicals in the tank, the water in the toilet tank (not the bowl) is drinkable.
- **Refrigerator:** Water can also be in the water valve of your refrigerator, waiting to be used for ice.

# Eat Your Water Intake

On an average people get around 20% of their daily water intake from food sources. However, in an emergency this would be an excellent source of pure water packed with rich nutrients.

## WATERMELON

Watermelon which is 92% water, is the best food for re-hydrating your body. It is also dense with vitamins and minerals, including vitamin C, potassium, magnesium.

## Fruits/Vegetables with Over 90 Percent Water Content

- watermelon
- coconut
- cucumber
- cantaloupe
- grapefruit
- strawberries
- broccoli
- cabbage
- cauliflower
- celery
- iceberg lettuce
- sweet peppers
- radishes
- spinach
- zucchini
- tomatoes

# 70 percent water content

- apples
- apricots
- banana
- blueberries
- cherries
- cranberries
- grapes
- oranges

- peaches
- pears
- pineapples
- plums
- raspberries
- carrots
- green peas

## CANNED FOOD

Bottled, canned juices, or canned food are an excellent source of water. Sealed food containers such as cans or bottles of juice, soup, fruits, vegetables, should contain liquids that are safe to drink.

# Plants That Contain Water

## HENS AND CHICKS

Hens and chicks, are little succulents that look like roses in bloom made of aloe leaves. The leaves of the plant are entirely edible and extremely hydrating. They taste sweet, crunchy and a bit astringent.

## GRAPE VINES

Grape vines are a wonderful source of water throughout the spring season. Small vines about ½ inch in diameter can be cut a few feet above ground and will drip water for hours. Larger vines will gush out water. Make absolute certain that you have a grape species and not something poisonous.

## BERRIES

Berries, wild or domestic, are full of water. But some are poisonous, so make sure you are aware of the difference. If they are the edible kind, you can eat the leaves and stems as well as the berries.

## MINER'S LETTUCE

(Claytonia perfoliate) is a wild plant grows throughout most of the US. Miner's Lettuce is extremely high in vitamin C and an excellent source of hydration. It starts growing in the spring and likes shady places.
Miner's lettuce is described as sweet and juicy. You can eat the leaves, stems and flowers of the plant.

## MINT

The stems of mint are always square in shape, and the leaves are serrated. They may also have pink or white flowers that can be ate. They're a good source of water and are dense in nutrition. Mint is high in vitamins A and C, thiamin, folic acid, riboflavin, manganese, magnesium, copper, potassium, iron, calcium and zinc.

## MAPLE AND BIRCH SAP

Sap consist mostly of water. Though sap mainly runs in the early spring, it's a great source of water and it's relatively easy to tap in to. You also get a significant yield and there are essential minerals in the sap that are good for you. If it isn't spring, you can still wring water out of the fiber of the trees themselves.

## CACTI

Prickly pears, or Opuntia cacti, grow pads and fruits that are completely edible. They are an incredible source of water. The fruits ripen in late September, they are bright red with a bright orange center. Others are yellow on the outside and green on the inside. You can eat them raw.

## VINES

Safe vines give a clear fluid, and poisonous ones are milky or yellow. Even the nonpoisonous ones can give you rashes if you put the leaf directly inside your mouth. So instead let the liquid drip into your mouth or use a container to catch the water.

## PALM TREES

Palms are the best ones to seek out for sources of fresh water. Simply bend one of the flowering stalks and cut its tip, then let the water drain from it. If you cut another thin slice off the stalk a couple of hours later, the flow will resume.

# Water Collection Methods

## 3 STEPS TO EXTRACT WATER FROM PLANTS
### Step 1: Tools
- Clear plastic bag with no tears
- String
- A plant

### Step 2: Choosing the plant
- The best types of plants have large, green leaves. Avoid toxic plants!
- Select a plant that receives the most amount of sunlight possible. The heat from the sun will speed the process.

## Step 3: The process
1. Choose a branch that has many healthy leaves.
2. Gently shake the branch to rid insects.
3. Place your plastic bag over it.
4. Tie the plastic bag tightly so no water escapes.
5. Make sure part of the bag hangs lower than the point where you tied the bag to the branch.
6. Water will run down the bag and collect at the lowest point.
7. Hang as many bags as you can for larger water quantities.
8. It will take approximately 1-4 hours for water to accumulate.

# Digging Your Own Well

Digging a water well is not complicated. It just takes a little time and a bit of sweat.

## HISTORY OF AN EARLY WATER WELL

Early wells were relatively shallow holes. These were placed near lakes, streams or springs. The purification or filtering was done by sand or simply the surrounding earth. Water that entered the well had to pass through the ground first. Some filtering, a little micro biotic action and you had clean, tasty water.

### Drawbacks to the shallow well

- They are susceptible to contamination from surface water.

- Only suitable for low water production.
- They dry up.

## Advantages of a shallow well
- Uses natural drainages, springs and high-water tables.
- Can provide a small family with enough drinking water.
- Can be dug by hand in the event of an emergency.

## Location of surface water
When you start to dig a shallow well, you are creating a low-pressure location for water to accumulate. This is the same way perforated drain pipes work. Water will be naturally drawn to a depression or hole because it is the path of least resistance.

## DIG A TEST HOLE
If you are going to dig test holes, wait one day and then check those locations for water. The holes should be at least 12" - 24" deep.

## Digging your well right
1. The location should be at least 36" deep and 36" or more across. You are trying to capture or draw water into the start of the well.
2. At this point you may notice a trickle of water entering your pit. [Mark that location!]

3. If the water's entry isn't obvious you will have to check your "digs" every hour for changes.
4. You are looking for the location (entrance) of water.
5. If the well just seems wet with no obvious entry point you will need to over dig the depth of the spot to create a holding reservoir.

## WATER IS ENTERING – WHAT DO I DO NEXT?

If the water enters from the side of the well.

1. Continue digging to a depth of 6 feet or more.

2. Shore up the sides of your well as you dig. You can use 2" x 6" or 2" x 8" lumber to temporarily hold the sides or walls back. Remember you are making a water reservoir to draw water from.

3. When you have reached the targeted depth, you will continue to over dig the center of the bottom. Then place bricks or connected flat stones as a base. It is a good idea to wash off these stones or bricks if you can.

4. Once the base is finished, carefully remove the bottom 2" x 6" or 2" x 8" braces. Some soil will sluff in. Using large river washed rocks sized from 6" to 18" build a circle which will form the first bottom row of your well casing. If possible, backfill behind the rocks with coarse gravel. This will support the

outer wall of the well and allow drainage water flow to fill the pit.

5.  Continue with this ring of large rocks until you reach the top of the well.

6.  It is preferred that you taper the rock wall casing. Have it smaller at the bottom and slightly larger at the top.

# Digging A Seep

Keep an eye out for damp or dark areas of dirt. This may signify ground water underneath. This is where you can dig a seep.

## HOW TO:

1. Dig wide (about 2 - 3 feet across) and about a foot deep. Your seep should hold a few gallons of water. As you dig watch for water to slowly seep into the hole.
2. Next, line the walls and the bottom of the hole neatly with small rocks.
3. This water should be safe to drink, as it's coming right out of the ground, filtered naturally by dirt.

4. Filter the water through a shirt or other cloth to remove sediment. If you have the means, boil it before drinking just to be absolutely sure it's safe to drink.
5. If you return to your seep later, wildlife may have consumed water from it, and it may no longer be clean.
6. To help reduce wildlife from drinking from the seep, look for large, flat rocks you can cover the hole with. Or drape tree branches over the hole, supported by two mounds of rocks on either side.

## DIGGING A SEEP IN A DRY STREAM BED
1. Follow a dry stream bed until it bends. Look for water on the outside bend.
2. As you're inspecting bends in a dry stream bed, look for dirt that is a shade darker than the surrounding dirt. Also, look for a shallow depression.
3. Start digging.
4. Once you've gotten down to mud, dig a little deeper, and put in a small container for collecting water.
5. Check back every 15 to 30 to check the amounts of water collecting in your seep.
6. If no water has appeared, you can try digging in a second location.

# Water Finding Tricks

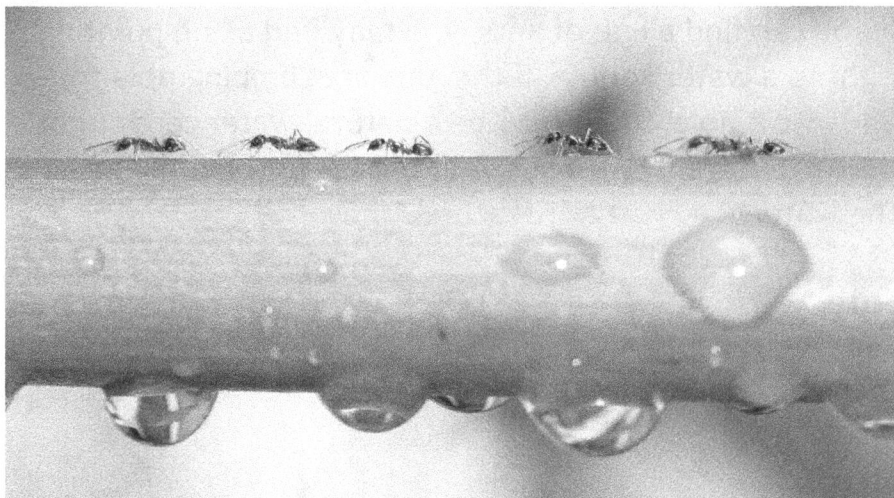

Watch for Signs of Wildlife to Signify Water.

## ANIMAL TRACKS

You'll also want to look for signs of wildlife, such as animal tracks in the ground. When animal tracks come together on game trails, often these trails will lead to water.

### Animal Feces

You may not spot tracks as easily as you spot animal feces. Consider heading in that direction to look for water.

### Mosquitoes

The presence of mosquitoes should indicate there is water nearby, usually standing or slow-moving bodies.

### Bees

Bee hives are located within 3 - 5 miles of a water source.

## Ants

If you can find a line of ants, you may find a line pointing right to a water source. If the ants are climbing up a tree and into a hole, there may be a natural water catch right inside that tree. Use a straw or dig a large hole to retrieve the water.

## Flies

Flies typically travel no further than 100 meters from a water source.

## Birds

Watch for small grain-eating birds and pigeons, specifically in the morning and evening hours. A bird's direct flight (one that is straight over land) can lead you in the direction of water.

# How to Clean Water

## BOILING

The most effective way to remove both viruses and bacteria from water is simply to boil it. Boiling will not evaporate all forms of chemical pollution, but it is still one of the safest methods of disinfection. Five minutes of a rolling boil will kill most organisms, but ten minutes is safer.

### Boiling can be done over
- campfire
- stove
- metal
- ceramic
- glass container

If no fireproof container is available, heat rocks for 30 minutes in the fire and place them into your container of water.

### This container could be
- rock depression
- bowl burned or carved out of wood
- folded bark container
- hide
- animal stomach

# Extreme Contamination

## COMMON WATER CONTAMINANTS AFTER A DISASTER:

- radiation
- lead
- salt
- heavy metals
- parasites
- harmful bacteria

# DISTILLATION

In a scenario where the only water available is dangerous water, there aren't many options. The safest solution is water distillation. Water can be heated into steam, and the steam can then be captured to create relatively clean water, despite its prior contamination.

Distillation won't remove all possible contaminants, like volatile oils and certain organic compounds, but most heavy particles will stay behind.

# SURVIVAL STRAW

The survival straw is a great device that is lightweight and easy to carry around. It is designed to disinfect and filter water. Most of these filters contain an activated carbon filter element, which not only filters out larger bacteria and pathogens, but also removes odd flavors and odors from the water.

# IODINE 2%

The most effective and affordable way to purify water is to add a couple drops of tincture of Iodine 2% to your water bottle.

> ➤ Make sure you're buying "tincture of Iodine 2%" not something else.

# IODINE ISSUES

> ➤ The toxicity and flavor of iodine can be a little problematic.

- ➤ Not a good choice for pregnant women.
- ➤ Or people with thyroid issues.
- ➤ Or those who suffer from shellfish allergies.
- ➤ Picky children do not like to drink iodine-infused water.

# BLEACH

Bleach can be carefully used to disinfect water with good results. The amount of bleach you will use depends on the water quality and temperature.

- ➤ Cold or murky water needs a little more disinfectant (four drops per quart)
- ➤ Warm or clear water (two drops).

## What to do next

1. After adding the chemical, put the lid back on your water container
2. Shake vigorously for a minute.
3. Then turn the bottle upside down.
4. Unscrew the cap a turn to let a small amount of water flow out to clean the bottle threads and cap.
5. Screw the lid back on tight and wipe the exterior of the bottle to get the chlorine on all surfaces.
6. Set the bottle in a shaded place for 30 minutes.
7. If the water is clear and at room temperature, then drink it.
8. If not, add another drop or two and wait another 30 minutes.

# UV Lights Devices

These devices purify water by treating it with UV light. They kill the viruses and bacteria but works best if you filter the water from heavy debris first.

## How does UV Light work?

UV light is very damaging to small organisms. When used as a disinfection method, it's very effective.

- It disrupts the DNA of illness-causing microbes.

# SUN

Solar water disinfection or SODIS, is a method that uses the sun's energy to disinfect contaminated water. The sun's abundant UV light kills or damages almost all biological hazards in the water.

## The advantages
- It's easy.
- Inexpensive or free.
- Offers good bacterial and viral disinfection.
- No dangerous chemicals.
- Does not require constant attention.

## The most common technique
- Lay out plastic bottles full of contaminated water in the sun for a minimum of one day. (2-liter bottles Maximum)
- Or 2 full days if the weather is overcast.

## Downfalls
- You need sunny weather or two days of overcast.
- You cannot do it in rain.
- It may be less effective against bacterial spores and cyst stages of some parasites.
- the water and the bottle need to be clear.
- The bottle should not be made of glass.
- This method does not treat chemical contamination.

- Only small bottles can be used. (2-liter bottles or less)

# Make Saltwater Safe to Drink

There are two ways to condense salt water.

## FIRST METHOD
1. Fill a metal container or a large pot with 2 - 4 gallons of saltwater.
2. Heat large rocks in a fire.
3. Once hot, drop the hot rocks into the pot of saltwater.
4. When the water is boiling, or producing vapor, cover with a towel, shirt, cloth, or even a blanket.
5. Use the cloth to absorb the water vapor.
6. Once damp, ring the water out of the cloth directly into your mouth or a water bottle.

## SECOND METHOD

1. Dig a narrow pit capable of holding 2 - 4 gallons of saltwater.
2. Heat up rocks in a fire.
3. Once hot, drop the rocks into the pit of saltwater,
4. When the water is boiling, or producing vapor, cover with a towel, shirt, cloth, or even a blanket.
5. Use the cloth to absorb the water vapor.
6. Once damp, ring the water out of the cloth directly into your mouth or a water bottle.

# Building a Solar Still

Building a solar still can help you get water when all other fast options are out of reach.

## How To:

1. dig a hole large enough to hold 5 - 10 soccer balls. Place a clean container, milk carton, or 2-liter soda bottle at the bottom of the hole. Cut the top off the container, if there isn't already a large mouth hole.

2. collect green vegetation and line the hole around the container.

3. If you have pond water or saltwater present, you can pour small amounts of water into the hole also, just enough to dampen the hole and vegetation lining it.

4. Be sure not to let this water or vegetation get into your container for capturing clean water.

5. Cover it with a tarp. Anchor each corner of the tarp with a large rock.

6. Place a rock dead-center of the tarp, right above the container that is at the bottom of the pit. This creates a drip point into the mouth of your container.

7. Every few hours replace the drying vegetation with additional green vegetation.

8. Again, pour a small amount of pond or saltwater into the hole; urinate into the hole.

## HOW IT WORKS

That container is there to collect condensation that forms on the underside of the tarp. The water vapor is produced by the heat inside the pit, forcing the water to condensate and purify itself.

# How to drink it

You will now need a way to drink any water that is collected.

**There are two options.**

1. Pull back the tarp, reach down and grab the container you've used and simply drink from it.
2. Use an extremely long straw that will stretch from the container at the bottom of the pit and out from under the tarp. Then you can drink the water without having to disrupt your water purification process.

## NO WATER AVAILABLE

If no sea or pond water is available, you can even urinate into the hole (just not into the container). Also, if you don't have a tarp, you can use a large garbage bag or even a water proof poncho.

# Building a Tripod Water Filter

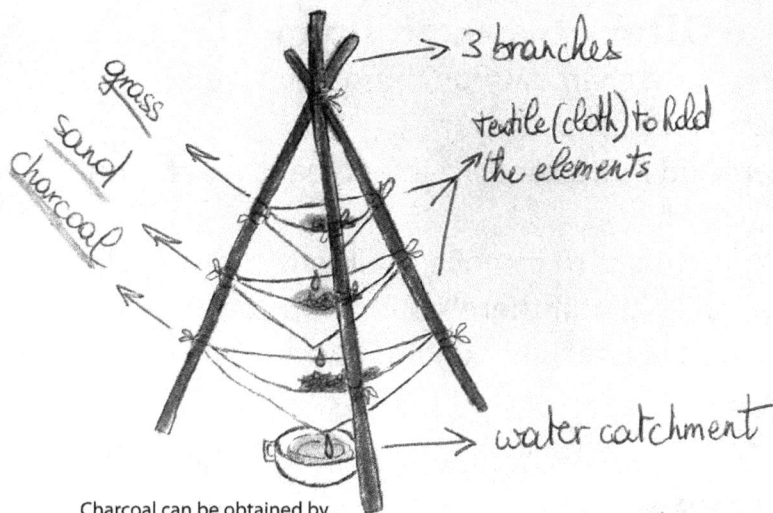

3 branches

grass
sand
charcoal

textile (cloth) to hold the elements

water catchment

Charcoal can be obtained by using burned firewood.

When you have pond or lake water that is not extremely contaminated, you can use this method to produce large quantities of drinking water.

## CREATE 3 LAYERS

1. You will need three long, sturdy sticks or poles.
2. You'll then use three sections of sheets, bandannas, or even shirts and create three layers,
3. Place the widest layer a few inches off the ground, tied to each of the three sticks.
4. A few inches above that tie the next layer.
5. Finally, a few inches above that, tie the top layer.

6. Place a large open mouth container at the very bottom.

## Create filters for each layer
**1$^{st}$ layer:** Place charcoal from burnt tree wood
**2$^{nd}$ layer:** Place sand and dirt
**3$^{rd}$ layer:** Add more sand, grass, or leafy vegetation

**Note:** Your goal is to remove sediment and particles in the water. Realize that if there's dangerous bacteria like Giardia in the water, this kind of filtering will not remove the bacteria.

## FILTERING WATER
1. Pour pond water onto the top layer and watch as it slowly drips down to the bottom layer.
2. have a container ready for capturing the water.
3. The final results won't look like clean water from the sink, but it will be a lot cleaner than what you first started with.

# Things to Avoid

## STAGNANT OR NOT?

It can be hard to tell if a body of water is stagnant or not.

### Here is a trick:

- Place a leaf in the water and see if the leaf drifts from movement in the water, or if it stays put. No movement signifying stagnant water that's not safe to drink unless filtered.

**FYI**: As long as the wind isn't blowing, this is a good way to tell if the pond is stagnant or not.

## AVOID MARSHES AND SWAMPS

typically, marshes and swamps carry a higher concentration of bacteria and parasites. These bodies of water are stagnant and should be avoided.

## SMALL LAKES AND PONDS IN AGRICULTURAL AREAS

Avoid water near agricultural areas. Illegal dumping and pesticide runoff will increase the chances of unsafe chemicals that are impossible to filter.

# Extending Clean Water Shelf Life

## SILVER

It is an extremely cost-effective way to keep water clean and drinkable for long periods of time.

### What to do

- Filter your drinking water.
- Submerge pure silver into the water.
- Your water will have an extended shelf life

### Why it works

- Pure silver releases a very low count of ions slowly over a period of time.
- These ions destroy harmful bacteria and organisms in the water, preventing them from multiplying.
- Silver works by interrupting the bacteria's ability to form chemical bonds. These bonds are needed for it to survive and reproduce.

# Creating A Fire

# How to Make Tinder

Tinder is made by using small, dried materials that are highly flammable. When you make a spark, you will need a tinder, to turn the spark into fire. Then you can transfer the small fire into a larger one.

## Materials found in nature suitable for a tinder

- Dry grass
- Leaves
- Shaved bark
- Dandelion head
- Birch bark
- Cattail fluff
- Cattail leaves dry
- Dry pine needles
- Fat wood
- Fungus
- Punk wood
- Poplar Cotton
- Jute twine

# Unusual Tinder Ideas

While I am a huge believer in knowing how to survive in the wilderness; it is more likely that you will be in your home or community when a catastrophe strikes. So, understanding how to use items that are commonly found in a home, is very important to your survival.

Below are tinder ideas that will allow you to quickly create a fire, for water purification, cooking and heat.

## CRUMPLED PAPER PRODUCTS:

Newspaper, paper towels, toilet paper, old notebooks, sticky notes, packaging, cupcake paper, coffee filters, and scrap paper.

## TRICK BIRTHDAY CANDLES:

Once you get this type of candle lit, it can be easily moved around without burning out. This will give you enough time to light other things on fire that couldn't be used as tinder.

## COTTON GAUZE:

Paper-wrapped cotton gauze is very flammable.
Tampons or Pads:
Cotton fiber feminine products when ripped apart make a great tinder. Also use the box and wrapper.

## PLANT-BASED CLOTH:

Any dry cotton, linen, or other plant-fiber clothing or cloth can be burned. Tear off strips to burn, rather than using the whole item at once.

## DRYER LINT:

This lint fluff is explosively flammable and works great in a pile of other tinder materials.

## WOOD SHAVINGS:

A small pile of wood shavings has a loose structure for combustion.

## GREASY CHIPS AND SNACKS:

A greasy chip bag is very flammable. Apply an open flame to the edge of any chip bag and watch it burn like a torch. Most chips you have in the cupboard will work.

## CARDBOARD:

Soak a little grease or oil into the cardboard for a better burn time.

## PLASTIC FIBER CLOTH AND ROPE:

Plastic ropes and cloth will burn when exposed to an open flame, if they haven't been treated with flame retardant chemicals.

## PETROLEUM JELLY COTTON BALLS:

Drench the cotton balls in petroleum jelly. The dry ones will burn for 20 seconds, but the petroleum cotton balls will burn up to 5 minutes or longer. You can also use this idea to increase the burn time on fabrics.

# Starting A Fire in the Wilderness

## SPINDLE & FIREBOARD BASICS

Friction-based fire making is more challenging but using the right wood will save you a lot of time and energy.

## SPINDLE

A spindle is the stick used to spin, to create the friction between it and the fireboard. If you create enough friction between the spindle and the fireboard, you can create an ember that can be used to start a fire.

### Best wood for fire board and spindle sets

- Cottonwood
- Juniper
- Aspen

- Willow
- Cedar
- Cypress
- Walnut

## Tips

➢ All wood must be bone dry. If the wood is wet, you will need to dry it out first or find something better.
➢ Avoid other moisture coming from the ground, or rain overhead.

# The Hand Drill

The hand drill method is the most time consuming. All you need is wood, a set of hands, and some serious determination.

## HOW TO:

1. Build a tinder
2. Cut a v-shaped notch into your fire board and make a small depression next to it.
3. Select a spindle, it should be about 2 feet long.
4. Place the bark underneath the notch. The bark will be used to catch the embers from the friction.
5. Place the spindle into the depression on your fire board.
6. Start spinning.
7. Maintain pressure on the board and start rolling the spindle between your hands and run them quickly down the spindle.
8. Keep doing this until an ember is formed on the fireboard.
9. Once you see a glowing ember, tap the fire board to drop the ember onto the piece of bark.
10. Transfer the bark to your nest of tinder.
11. Gently blow on it to start your flame.

# Bow Drill

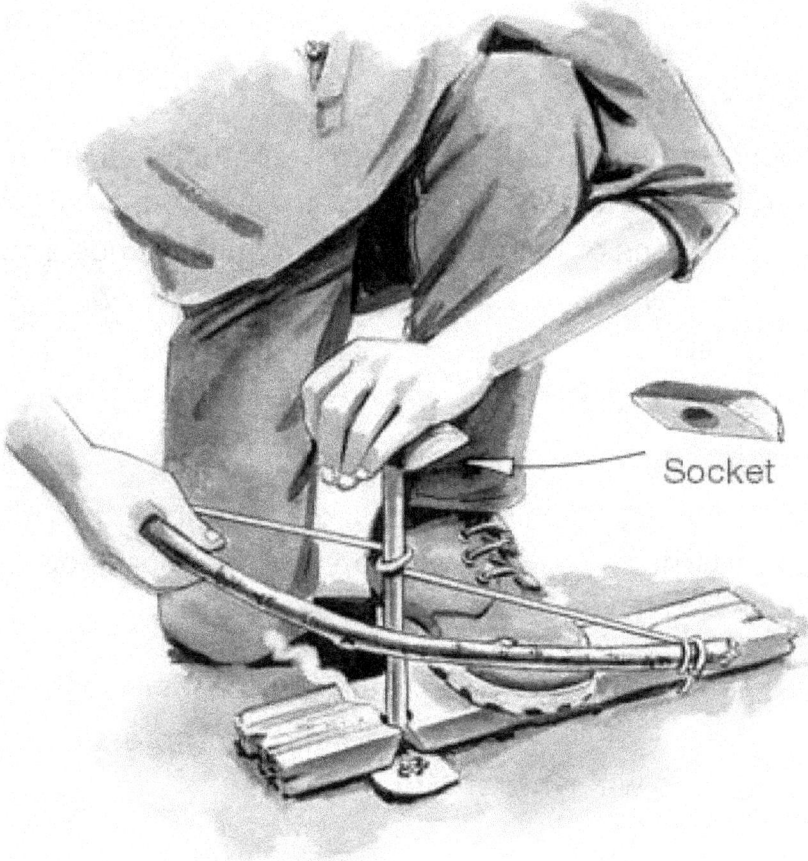

Socket

The bow drill is probably the most effective friction-based method to use, because it's easier to maintain the speed and pressure. In addition to the spindle and fireboard, you'll also need a socket and a bow.

Find a socket

A socket is used to put pressure on the other end of the spindle as you're rotating it with the bow.

- The socket can be a stone or another piece of harder wood.

## MAKE YOUR BOW

The bow should be close to an arm's length. Use a flexible piece of wood that has a slight curve.

- The string of the bow can be anything. A shoelace, twine or a rope. Just find something that won't break.

## PREPARE THE FIREBOARD

Cut a v-shaped notch and create a depression next to it in the fireboard. Underneath the notch, place your tinder. String up the spindle

Catch the spindle in a loop of the bow string. Place one end of the spindle in the fireboard and apply pressure on the other end with your socket.

## SAWING TO CREATE A SPARK

Using your bow, start sawing back and forth quickly. Keep sawing until you create an ember.

## MAKE A FIRE

Drop the ember into the tinder and gently blow.
Flint and Steel

Starting a fire by flint and steel is an old reliable method. If you find yourself without a flint and steel set, you can always improvise.

## SUBSTITUTES

- Quartzite for the flint.
- Steel blade of your pocket knife for the steel.
- Fungus or birch for the char.
- Any solid rock will work.

## How to:

1. Grip the rock and char cloth.
2. Take hold of the piece of rock between your thumb and forefinger. Make sure an edge is hanging out about 2 or 3 inches.
3. Grasp the char between your thumb and the flint.
4. Grasp the back of the steel striker or use the back of your knife blade.
5. Strike the steel against the flint several times.
6. Sparks from the steel will land on the char cloth, causing a glow.
7. Fold up your char cloth and put it into the tinder.
8. Gently blow on the tinder to start your fire.

# Lens-Based Methods

Using a lens to start a fire is an easy matchless method. To create a fire, all you need is some sort of lens to focus a beam of sunlight on a specific spot.

## EASY TO FIND MAGNIFIERS

- Magnifying glass
- Eyeglasses
- Binocular lenses

## How to:

1. Angle the lens towards the sun to focus the beam at your tinder.

2. Put your tinder under this spot and it will start to smoke.
3. Gently blow your tinder to start the fire.

## Drawbacks

- The lens-based method only works when you have sun.

# Fire from Ice

All you need to do is form the ice into a lens shape and then use it as you would when starting a fire with any other lens. This method can be particularly handy for wintertime camping.

## GET CLEAR WATER
For this method to work, the ice must be clear.

### Water sources
- Tap or bottled water
- Use clear lake
- pond water
- melted snow.

## SHAPE YOUR ICE BLOCK
The best way to make a magnifying glass is to fill up a bowl, cup, or a container made from foil. Or in a real pinch, find a larger rock that has a relatively smooth indentation. This should look like a bowl. Clean it off well and pour the water on to the rock. Let the water freeze and pop the ice out.

## LET THE WATER FREEZE
Your block of frozen water should be around 2 inches thick for best results.

## FORM YOUR LENS

Use your knife to shape the ice into a lens. The lens should be thicker in the middle and narrower near the edges.

## POLISH YOUR LENS
Polishing it with your bare hands. The heat from your hands will melt the ice so that you form a smooth, shiny surface.

## START A FIRE
Angle your lens towards the sun. Focus the light beam on your tinder to start the fire.

## AN ORANGE
Citrus oils are flammable. If you can get some sparks into an orange, you can set it on fire.

**How to:**
1.   Cut a hole in the top of an orange.
2.   Clear out some of the flesh.
3.   Let the orange dry for a few hours.
4.   Shove a hard rock down the middle of the orange.
5.   Place a knife or stick down the center of the orange on to the rock.
6.   Rub the tool rapidly to cause the rock to spark.
7.   The entire orange will catch on fire, so transfer to your tinder quickly.

# Starting A Fire in Urban Areas

# Disaster Can Happen When You're Not Prepared

Every survivalist knows that you are supposed to load up on fire-starting materials. Having a way to cook food, purify water and keep warm during a disaster is crucial.

However, you could find yourself in a situation where you don't have access to your supplies. If that happens, you will have to use whatever materials you can find. Fortunately, there are many ways to start a fire without matches, flint, or a lighter.

These methods may be bizarre, but they all work, and they work well. Be sure to always use safety precautions when handling fire. In addition, make sure to properly extinguish one that has been lit.

# *Coke Can and Chocolate Bar*

## WHAT YOU WILL NEED
- 1 soda can
- 1 chunk of chocolate
- Sunny day

## How to:
1.    Polish the bottom of the soda can with the chocolate. Do this by rubbing the chocolate on the

bottom of the soda can. It will make the bottom of the can shine like a mirror.

2. Sunlight will reflect off the bottom of the can, forming a single focal point.

3. Place the tinder about an inch from the reflecting light.

4. Aim the beam directly at your tinder.

5. Shortly, your tinder will catch on fire.

# Batteries and Steel Wool

## WHAT YOU WILL NEED

- Steel Wool
- 9-volt battery (or whatever is handy)

## How to:

1. Stretch out the steel wool. You want it to be around 6 inches long and a ½-inch wide.

2. Hold the steel wool in one hand and the battery in the other. Any battery will do, but 9-volt batteries works best.

3. Rub the side of the battery with the "contacts" on the wool.

4. The wool will begin to glow and burn. Gently blow on it and transfer the wool to your tinder.

# Gum Wrapper and Battery

## WHAT YOU WILL NEED

- 1 battery
- Foil gum wrapper

## How to:

1. Use any size cylindrical battery.
2. Cut or rip the foil wrapper so that the middle is about 1/4 as thick as the ends.
3. Touch the foil side to the bottom of the battery, then touch the other end to the top of the battery.
4. The narrow part in the middle will catch fire.
5. Quickly move the fire to your tinder.

# Jumper Cables and a Car Battery

This method is a last case scenario since, you risk blowing up your car battery. But in a pinch, it may be worth a try to avoid freezing to death.

## How to:

1.  Attach a set of jumper cables to your car battery.
2.  Take the opposite ends and tap them together to create sparks.
3.  Transfer the spark to your tinder.

# Pencil and a Car Battery

If you don't want to make sparks with your jumper cables and risk ruining your car battery try this instead.

**How to:**
1.  Carve down **two ends** of a pencil until the lead is exposed.
2.  Attach your jumper cables to both ends of the pencil, making sure the jumper cables are touching the lead.
3.  Place the pencil on your tinder.
4.  Then attach the other ends of the jumper cables to your car battery.
5.  The pencil will heat up and set your tinder on fire.

# Batteries & Cell Phone Charger

## WHAT YOU WILL NEED
- 2 D-batteries
- Cell phone charger

**How to:**
1.  Strip your charger cord until the ends of the wires are exposed.
2.  Hold two D batteries together, press one end of the cord to the bottom of one battery, then touch the

wires on the other end of the cord to the top of the other battery.
3.    Have some tinder touching the top of the other battery also.
4.    The heat will cause the tinder to catch fire.

# A Condom Filled with Water

## WHAT YOU WILL NEED
- 1 condom
- Water
- Sunny day

## How to:
1.    Fill a condom with water until it's the size of a small water balloon.
2.    Then hold it in the sun at an angle.
3.    Focus the beam of sunlight onto your tinder.
4.    Within a short time, your tinder will begin to smoke.
5.    Gently blow on it to get the fire going.

# Urine and a Sandwich Bag

## WHAT YOU WILL NEED

- 1 clear sandwich bag
- Urine
- Sunny day

## How to:

1. Fill a clean plastic bag with urine.
2. Seal it.
3. Use the urine-filled bag like a magnifying glass.
4. Focus the beam of sunlight on your tinder.
5. It will catch fire.

# Old TV Screen

If you have or can find an old big screen TV, take it apart and remove the giant Fresnel lens.

## How to:

1. Angle the lens so a beam of sunlight is focused on your tinder.
2. This method can actual boil a bottle of water. So, don't stick your hand in the sunbeam!

# A Flashlight

## WHAT YOU WILL NEED
- 1 flashlight
- tinder
- Sunny day

## How to:
1. Remove the top lens and pull out the reflective cone that the light bulb sits in.
2. Place a small tinder where the bulb would normally sit.
3. Put the cone in direct sunlight.
4. The reflection of the sunlight will spark the tinder.

# An Empty Lighter

An empty lighter may seem useless, but you can still use it to start a fire.

## How to:
1. Slowly grind the wheel of the lighter against a sheet of paper.
2. It will begin to accumulate a small pile of flint dust.
3. Place a small tinder on the dust.
4. Use the lighter to spark the dust.
5. The dust will catch the spark and set your tinder on fire.

# Brake Fluid and Chlorine

Wear safety goggles or glasses to protect your eyes. This method should be performed on the ground instead of in a container. There is no way to determine how the material in the container will react.

**How to:**
1.  Start with a small pile of powdered chlorine.
2.  Pour a small amount of brake fluid on it, then quickly back away.
3.  After smoking for a moment or two, the pile will burst into flames.
4.  Now place some tinder and wood on the pile before the flames go out.

# Baking Soda TP Roll

## WHAT YOU WILL NEED
- Non-melting lid, or a concrete bottom
- Toilet paper roll
- Tinder
- Water
- Baking soda

**How to:**
1.  You will need a large lid, or a non-melting base for this fire.

2. Place one sheet of toilet paper down on the base.
3. Place a small pile of baking soda onto the toilet paper sheet.
4. Cut a toilet paper roll in half, and then cut a ½ circle out on one side.
5. Stuff the roll with a few more sheets of toilet paper.
6. Sit the roll right side up over the baking soda with the half circle exposing the powder.
7. Place a pile of tinder over top of the entire roll.
8. Pour 1 Tbsp. of water onto the base.
9. Move out of the way, the fire will lite fast.

# Make A Magnifying Glass

# *Lightbulb*

## WHAT YOU WILL NEED

- Clear light bulb
- Water
- Balloon or plug
- Sunny day

### How to:

1. Chisel out a lite bulb at the bottom.
2. Once the bulb is open and you can see into the glass part, fill with water, and shake to clean out the white film.
3. Then fill with clear water.
4. Now seal the bottom closed if possible. (could use a balloon or a plug)
5. Put the new bulb-magnifying glass in the sun and point the rays at your tinder.
6. Within a short period of time your tinder will catch on fire.

# *Plastic Wrap Balloon*

## WHAT YOU WILL NEED

- Small bowel
- Water
- Plastic wrap

- Rubber band or string
- Dark paper (could be spray painted, if dry)
- Sunny day

## How to:

1. Get a small bowel.
2. Place a large piece of plastic wrap the top.
3. Pour water over the top until it sinks into the bowel.
4. Gather the edges of the plastic wrap and make a balloon shape.
5. Twist the plastic wrap balloon until it is tight.
6. Use a rubber band or string to secure.
7. Put the new plastic wrap balloon-magnifying glass in the sun and point the rays at a dark piece of paper.
8. The paper will catch on fire.

# *Picture Frame Plastic*

## WHAT YOU WILL NEED

- Picture frame (with glass)
- Plastic wrap
- 2 sturdy objects (concrete blocks, bricks, horse etc.)
- Warm or sun heated water
- Sunny day

## How to:

1. Remove the glass from a picture frame.
2. Wrap the entire frame with plastic wrap.
3. Make a structure using two objects to hold the frame above the ground on each end. (two blocks, bricks or anything handy)
4. Do not make a table that blocks the center.
5. Place the plastic wrapped frame on the above ground structure.
6. Pour warm water directly on to the frames plastic wrap.
7. The plastic will bow downwards making a perfect magnifying glass.
8. Find where the ray of sun is reflecting and place your tinder directly underneath.
9. Your tinder will catch on fire.

# Why You Should Read This Book

As I believe you will discover as you begin to read it, I am probably as close to the philosopher-economist from Mars as anyone you are ever likely to encounter anywhere on this little green planet.  As such, I am able to present you what I think is the most intellectually detached, objective and analytic view of what goes on around us that is currently available in print.

I am an American and was raised as part of a family which was engaged, through my Father's diplomatic service, in representing the richest and most powerful country on earth in a large number of other places of the world.  But I was born – and my Mother was born – in Nicaragua, one of the poorest and perhaps weakest, but also one of the proudest, most joyful and most poetic countries on earth.  By the time I was ten years old, I had lived in four countries, had traveled to at least four more, and spoke five languages well, including Arabic, French and Dutch.

The rest of my life has followed more or less on the same course.  As an international consulting economist I have now worked in well over two dozen countries around the world, rich and poor, weak and powerful, in North, Central and South America, Europe, Africa, South, East and Southeast Asia.  I have known Presidents and Vice Presidents, none too intimately of course, and served as direct advisor to Ministers of Economy and Finance in numerous places and times.  I have worked with and for large corporations and their industry associations, as also with and for struggling micro-enterprises and self-help community groups.

I have seen a lot since being delivered by the mothership, and have thought hard about everything that I have seen.  In the process, I have had to abandon many of my formerly most cherished beliefs, and face up to the fact that much of what we are encouraged to believe about ourselves in the world today – as citizens of so-called "advanced" nations, as enlightened professionals, as business leaders, as consumers, and as ordinary human beings – is simply false.  What is happening in this world, and what our role is in what is happening, is far more convoluted and complex than we are ever told on television.  There is a great deal of evil and destruction taking place, that we try not to let too deeply into our psyches, despite or maybe because of the deluge of so-called information we receive.

What we are almost never told, and never want to hear, is that we ourselves are responsible for much of what is happening. On many levels.

One, we could certainly almost all of us already be doing much more about what is happening, to make it somehow better. But to do so effectively, we first would need to understand what is happening better.

That is part of what this book will help you do.

Second, we are almost never told – though most of us unconsciously know – how much the more fortunate of us actually benefit from some of the evils we see all around us. Poverty, for example, acts as a cushion for the middle- and upper classes, providing a plentiful source of low-cost labor, for example. The rape of our environment produces cheaper goods for the current generation, whose full costs will only have to be absorbed by successive unborn generations in the future, who currently neither vote nor spend. We benefit from many of the "bads" that are produced by our system, and many of us unconsciously fear any change that might also reduce the "goods" that we are consequently able to consume.

These "bads" are neither "deliberate", nor "accidental". They are part and parcel of how the system works, but were not necessarily instituted by any specific individual or group of individuals, and do not necessarily reflect any original evil or selfish intent. These "bads" perpetuate themselves in time, however, through the operation of processes within our system that fundamentally exist to serve the needs and wishes of the most powerful among us – but also produce poverty, environmental destruction, violence and social alienation – not necessarily as the conscious wish of anybody alive or dead, but simply as a by-product of the workings of the system.

The problem that we must all now face, however reluctantly, is that the magnitude of the "bads" that are now being produced by our system is, for the first time in history, so great as to literally threaten to overwhelm and destroy the system itself.

# I:  Introduction – What System?

## *What "System"?*

The global economic and political system prevailing in the world today.

Having in large measure supplanted more traditional religious, clan- and class-based institutions in the West, it consists fundamentally of a secular, materialistic "free-market"-based system of production, along with the commercial, communications and security apparatus needed to sustain it.

## *Why is it "Broken"?*

Any system which not only reproduces but regularly and progressively deepens and extends systemic evils such as the following is no longer functional, and must be restructured in order to avoid multiple and imminent catastrophic failures.

The multiple evils that surround us include:

- large-scale poverty indicated by increasing numbers of people in the world who suffer child mortality, malnutrition, disease, ignorance, joblessness and social disintegration
- progressively deepening environmental degradation and the destruction of basic our natural resources of air, soil, water and all the living things that share the biosphere with us
- alienation, criminality, violence and insecurity at all levels of social existence:  from within what little is left of crumbling family structures; to the streets, neighborhoods and communities of our large cities; to the proliferation everywhere of institutional violence, systematically, blindly and soullessly carried out as much by nation states as by drug "lords", white slavers and other international gangsters, and by ethnic and religious haters in every corner of the modern world

While some apparent relative progress may be observed at times in some localities, the general trend is of accelerating deterioration on all fronts. The system is in crisis and will predictably produce catastrophic collapse, probably within a generation, unless steps are taken soon to admit, address and "fix" the fundamental causes of its failure.

## Can the System be "Fixed", or Is It Already Too Late?

This is without any doubt The Question.

History may seem to indicate that it can not be fixed by ordinary mortals simply making up their minds to change. Do human beings only react cohesively and effectively when forced to do so by cataclysmic events that periodically ravage and destroy the existing *status quo?* If so, then it probably is too late for us because the magnitude of upcoming catastrophic events – undoubtedly many times the size of any previously experienced, including such calamities as the world wars, revolutions and mass exterminations of the last century – may very likely this time cause irreversible damage. If not Armageddon, then at least something so close in kind and in scope as to make the future scarcely livable for surviving generations.

On the other hand, there are many things that are different today than they have ever been before:

- We *know*, or at least are capable of knowing, what is happening everywhere in the world, almost as it is happening, minute by minute
- We have *technologies* that, properly applied, give us the power, for the first time in history, to seriously address each and every one of the major systemic failures that beleaguer humanity today
- We have the *wealth* necessary to finance the investments needed to provide all people living on this earth with the basics of a dignified, productive and happy life, to preserve our common natural resources and to rescue our human civilization and our culture from the promulgators of evil that, by neglect, have been allowed to proliferate among us

*Knowing* what is going on, and having the ability to make a difference also makes us fully *responsible*, for the first time in history, for the evils that we can stop but will not.

This book will attempt to make clear and irrefutable what we already know to be true.  **The system _is_ broken**.  **We _must_ fix it**.  The only question is:  do we have the intelligence, the moral fortitude, the heart, and especially the _will,_ to do what has to be done before it is too late?

As to _what_ has to be done, the full answer will only become apparent to us as we come to grips with the reality of our urgent need to act, as we make the necessary personal and collective decisions, and as we begin to progress together upon the creative path of voluntary rational change that lies before us. No one person, and certainly not this author, can provide all the answers.  But this author and this book will at least attempt to push forward an exploration of the alternatives we may not all yet realize that we have, and point out at least a few of the obvious steps that – depending primarily on changes we can make within ourselves by the exercise of will – can be taken in the short term, say the next five years at most.  Beyond that we can also point to fundamental institutional and political reforms that should begin to be explored and discussed together now, so that we can build a consensus for peaceful change that can begin within a decade.  Some of these reforms may appear to be radical and risky.  But, we will argue, far less radical and risky than allowing the decay of existing world economic and political institutions to continue unchecked.

If, as intelligent and loving human beings – humbled by and grateful for the gift of life that we have miraculously received – we can use our knowledge, technology and wealth to create the means to communicate openly, argue rationally and peacefully, and govern our individual and collective behaviors so as to prevent the most heinous evils from befalling any of us while bringing about the most good for the most people, I am convinced that the work of reforming our institutions and societies and building the foundations for a world that is just, healthy and happy can be accomplished in no more than one generation.

We, the living on Earth in the year 2008, are, I believe, the generation that has been called upon to wake up History, recognize what is going on around us and, for the first time, take responsibility and control for the future of life on Earth.  We

7

can, I believe if we start now, create the "Heaven" on Earth that has been promised to us by our Creator and that depends now only on the realization of the promise that lies within us all.

# II.   What is the "System" And How Does It Work?

Because we, like fish swimming in water, are so deeply immersed in the system and its workings, we are not always very much aware of what it is or how it works.  Even, perhaps, that it is there at all.

So the first thing this book must do, as simply and as briefly as possible, is to point out the existence of a system (of systems) within which we all live, and how this system, in response to the needs of its "constituents" and the relative power among these, naturally and without deliberate guidance of any kind, continually regenerates and reproduces itself in time and space, exhibiting a deep inertial resistance to changes in either structure or direction.  Like any organism that "lives", this simply happens without the necessity for any individual or group to consciously direct the process, either wholly or in part.  The system breathes and metabolizes, cells are repaired or reproduced, blood circulates, signals are generated and transmitted – couples and communities are formed and the system perpetuates itself in space and time – without any apparent overall control.

Economic and political systems behave in a similar manner, tending to mobilize and direct resources and energy in ways that preserve and strengthen the power of the system to perpetuate itself for the benefit of its constituents, without, necessarily, any apparent overall control.

Obviously, some systems are more efficient than others.  Some do not work well from the beginning, and rapidly die out.  Others work for a long time, but perhaps fail to evolve in response to changing environmental conditions, and gradually die out, or are destroyed by competing systems that have evolved in a more adaptive manner.

## Our System Is Not Just Broken, It Was Defective from the Start: Still, It Can Be Fixed

Any system that has survived for any length of time, such as the materialist-productionist economic and political system that – in greater or lesser degree – has spread everywhere throughout the world since the early 19<sup>th</sup> century Industrial Revolution, evidently must have provided benefits to a large number of its constituents.  Our current system, perhaps epitomized by the US domestic political economy, but also fundamentally characteristic of the Chinese system of state capitalism and becoming daily more integrated on a global scale, continues to produce a great many benefits for a large number of its constituents, and this is what gives it the strength to continue to expand and proliferate.  The system also produces a great number of evils, however, which were enumerated above.

The fundamental point is that *it is the same system* that produces both the benefits and the evils.  The evils produced by the system continue to exist because they themselves may serve some purpose, and may provide substantial benefits, to other, more powerful, constituents of the system.
The second preliminary task to be accomplished is, then, to show how it is that the evils of poverty, environmental destruction, criminality and violence are themselves intrinsic to the functioning of the current system, how they bring benefits to its most powerful constituents – without necessarily any deliberate action or even awareness on their part – and how, therefore, they are also reproduced and perpetuated in space and time even though they threaten to ultimately destroy us all.

To put it crudely, "shit happens".  It happens every day, and all around us.  Each of us contributes, independently of our individual volition.  It's just part of the "system", and so long as it was easy enough to simply flush away, that was Ok.

Today, however, the oceans of our ignorance have been filled, there is nowhere left to flush it and all the other myriad forms of waste we "innocently" generate, and unless we take responsibility and do something to eliminate or transform it soon, we will end up drowning in it, probably sooner rather than later.  The system, which has apparently worked so well until now, must be changed fundamentally and soon, or it will end up choking us all to death.

## Flaws in the System

The realization that real world markets are imperfect – despite the obvious elegance of the theory of perfect competition – is nothing new. Even theoretical economics has gone a long way to explore and explain how imperfect information, high transactions costs and externalities of various kinds produce a myriad of "market failures" in the real world. As has been realized and advocated for decades by economists and politicians in the Keynesian mode – from Keynes[1] himself to Franklin Roosevelt to John Kenneth Galbraith in the 1960s – the undeniable and inevitable presence of market failures, and the destructive extremes which unfettered self-interest had been seen to produce if left entirely to its own devices – makes it necessary for there to be a "countervailing power", exercised by good government in the public interest and for the common good. Their view was, that is, that neither unchecked capitalism – as in the age of the robber barons – nor unchecked statism of either the right or of the left could protect the public interest from inevitable abuses. Much as is reflected in the U.S. constitutional conception as regards the ideal structure of government, Keynes, Roosevelt and their followers believed strongly that a strong and effective system of "checks and balances" had to be maintained and used in the public interest. President Eisenhower himself warned of the dangers of allowing power to become excessively concentrated in any one segment of society, including the "military-industrial complex"[2] from which he himself had sprung.

But the murder of President Kennedy, and the failure of the Johnson Administration, led ultimately to the election of Richard Nixon and the gradual emergence of a so-called neo-conservatism which has become the almost unquestioned orthodoxy of our times. "The market rules." The less

---

[1] John Maynard Keynes, celebrated British economist who advocated the management of government spending to compensate for demand fluctuations in the private economy, specifically during the Great Depression of the 1930s.
[2] In the penultimate draft of Eisenhower's 1960 Farewell Speech, he referred to the "military-industrial-*congressional*" complex, a fact which gives insight into how, in his mind, the collusion of government, paid for by money, makes it possible to keep the system in place and functioning.

government, the better.  The consumer is sovereign, and, whatever outcome the "market" produces is better, and represents more good for a larger number of people, than any other possible outcome.  It has to be so, because, they say, the consumer *is* sovereign and whatever outcome is produced is the result of the millions of free choices made by them in millions of transactions taking place everyday in the marketplace.  No one controls it.  What could possibly be better, or more *democratic*, than that?

## Untrue Myths about the Market, Freedom and Capitalism

In the following sections we will argue that, in fact, the consumer is not sovereign, and could not be sovereign even under the idealized conditions stipulated by economic dogma.  The system *is* controlled, but it is not controlled in the *public, or general* interest.  That is evident from the results that it produces.

Also, it will be shown how, under the current system, control is exercised by the few, almost entirely to serve their own selfish interests, and how certain structural features of the system allow them to continue doing so.  An understanding of how power is manipulated in the economic and political system in which we live today, may lead to insights as to how things might be changed to bring about a more balanced, more equitable and, ultimately, a more survivable situation for ourselves and our children.

### *The Fatal Flaw of the Free Market Model*

There can be little doubt that the competitive, free enterprise system such as evolved in England and Western Europe and later reached its culmination in the United States is by far the most effective engine to minimize costs and maximize innovation that has ever been devised by humankind.  Within a legal framework to define and protect property rights and to support the enforcement of contracts, and with certain care taken to facilitate entry and prevent anticompetitive collusive or predatory behavior, this system – based fundamentally on individual freedom and initiative – has shown itself to be the most efficient productive system so far encountered in history, and indeed a sound basis on which to organize the supply side of an economy.

Please note, however, that even on the supply side, the market depends fundamentally on certain institutional elements being present that, in the real world, may not be present or may not be allowed to function as freely and aggressively as may be needed to prevent the fraud, collusion and abuse that undermine competition and the market's ability to produce efficient, innovative production systems for the benefit of society.

However, on the supply side it is at least true in principle that great cost savings and efficiencies can be achieved.

Despite the best efforts of some of history's leading economists to provide a theoretical basis to close the system and provide intellectual support for neo-conservative *laisser-faire* approaches to economic management, it can be shown that where the market system fails completely is on the demand side, and in particular in determining – in any theoretically defensible way that is consistent with a liberal political ideology – *what* society's resources and productive apparatus *should* be organized to produce. It's the old guns vs. butter (maybe better, Las Vegas getaways vs. affordable medical care) conundrum.

Granted, once a list of the final demands for a society is specified, the most efficient way of organizing and managing production of this list of goods and services is through a free market system. But, how does the market determine the content of society's list of final demands in an optimal way? How can it determine how much of society's resources should go into Las Vegas casinos instead of community medical centers?

The answer is, it can't.

### The Myth of Consumer "Sovereignty"

Much work has been done, and Nobel prizes have been awarded for the construction and embellishment of the notion of "consumer sovereignty" which states, roughly, that consumers – through their individual choices in the marketplace – supported by their willingness to exchange tangible wealth in support of their consumption choices, ultimately determine what is demanded in society and therefore, what ought to be produced. "Equilibrium" and the highest level of welfare achievable by that particular society – given the specific preferences of the individuals that compose it – is achieved when supply equals demand, just satisfying the totality of their freely expressed needs and desires, at the lowest possible cost to society in terms of resource usage.

There are two fundamental problems with this characterization of the demand side of a market-based economic system.

First is the presumption that individuals somehow freely and independently can come up with their own individual lists of final demands, ranked in order of preference given the relative prices of the goods and services on their lists and the total income available to each. The individual consumer is presumed to have full knowledge of all the possible goods and services that they *could have*, if they so desired, and their relative prices. Also, they have a well-developed sense of the strength of their preferences for one item on their list of potential demands over any other, so that, in the event of changes in the relative prices of goods and services (through the interplay of supply and demand in the marketplace) they can immediately say how much their demand of one item would be reduced and how much this reduced consumption would be replaced by increased consumption of some other item(s).

Paul Samuelson, a well-known American economist of a generation ago, did a lot of work as a young man which later won him the Nobel Prize in Economics in developing the so-called *theory of revealed preferences*, which argued that, even if people were not actually able to articulate a well-ordered list of their final demands and the relative values they placed on each item of the list, one could, by observing their behavior as consumers in a free market, determine their implicit preferences as revealed by their actions. Since, by assumption, these individuals were acting in full knowledge of their possibilities and free of any external influence or constraint, and, since again by assumption, the ultimate list of goods and services produced and consumed within the economy as a whole was simply the aggregation of all of the individual consumers' lists of final demands, there could not possibly – within a *free* market system

14

– be any way to improve on the final outcome without denying at least some people the full satisfaction of their needs and desires.

This is obvious, though perhaps elegant and sophisticated, theoretical nonsense. As is evident both from introspection and from observation of the behavior of real people all around us, *none of us* comes into this world with anything remotely approaching full knowledge of the possibilities that could be brought to lie before us, even with already available technology, much less by investment in the development of new technologies that could open new possibilities for us as individuals and as societies. We, in fact, are *taught* what it is we ought to like and what it is we ought not to like, by a whole host of influences – parents, school, church, community, culture, commercials and "infomercials" –each having diverging notions of what we could have and what we ought to want.

The corporate world is of course deeply interested in teaching us to want their products, and yearly spends hundreds of billions of dollars in direct advertising that spews its messages across the airwaves, in print, and even on the sides of buildings in cities across the world. Their commercial interests and the messages they transmit are not limited to direct commercial advertising, however. As owners of major communications conglomerates, major corporations have a huge influence on the messages that are transmitted through entertainment products of all kinds, and even through the selection of the kind of so-called information products that we are presented. Corporate involvement in politics at all levels, also helps to determine the kinds of messages that are transmitted by individual politicians as by public institutions, helping to form in the only semi-conscious awareness of individuals their notions as to what they can have and what they should want.

Thus, in practice, the final results generated by the presumed interaction of supply and demand in a market economy are heavily – if not yet exclusively – determined by deliberate actions taken by large corporate interests who dominate the supply side of the economy. Producers no longer respond to the autonomous demands of consumers. Rather, by their massive and all-pervasive efforts to shape consumer preferences, they act to condition consumers into wanting to buy the products which are more profitable for themselves. The market, in short, does not guarantee an outcome that is necessarily either in the individual nor in the general public interest.

How could it be either in the individual or the public interest that the market in the U.S. has produced a medical/pharmaceutical industry that is no longer oriented either towards the prevention or the curing of disease, but rather to the extension of life – under heavy, hugely expensive and hugely profitable medication –among a populace made and kept chronically ill by ignorance, neglect and an increasingly toxic environment? Research and development, even government-sponsored, tax-financed R&D, is now almost wholly focused on the development of drugs that will keep the chronically-ill alive, not to teaching prevention nor to finding cures. Once the models of disinterested, self-sacrificing, almost saintly researchers and healers devoted to alleviating human suffering, doctors have now come to be viewed as little more than highly-educated social parasites, interested first, second and third in making money for themselves by providing only the bare necessary minimum of care and attention to the needs of their patients.

How could it be that the new "rich" among us would prefer to buy a two million-dollar house, when they and their families could live comfortably and safely in a half-million dollar house, with the remainder of their expenditure used to finance or guarantee the down-payments for an additional 30 or more homes for lower-income families?

How can the Japanese epicure be so conditioned to extravagance that he is willing to have 1,000 endangered whales sacrificed annually just to satisfy his appetites, and get his government to sanction his hedonistic cravings through the blatant abuse of their public trust and the name of science?

How can our young prefer an orgiastic lost weekend in Vegas or Cancún to a few days visiting a Navajo reservation and exploring ancient American cultures and the beauties of the American Southwest?

How? Because they are taught and conditioned at great expense to develop these values and these tastes, and to use whatever money they may have to exhibit these behaviors, always, you may be sure, for somebody's profit.

The important issue is to realize that the values, tastes and preferences that guide the consumption decisions of individuals in a free market economy are taught.  They are not in any but the most simplistic ways inherent or natural to the human condition.

Once we realize that values, tastes and preferences are taught – and that there are many interest groups competing to determine the specific contents that are taught – then we can begin to think about what *ought* to be taught, in the interest of bringing into being a just, healthy, humane and happy society for ourselves and for our children.  We can then also see through and roundly reject the blatantly false message of free market ideologues that any deliberate interference – whether by government acting in the public interest or by community or religious groups acting in furtherance of their own beliefs – will only produce "distortions" and infringe on the God-given "liberties" of right-thinking Americans.  That is not true.  Rather, the corporate apologists of a false and distorted conception of the free market have almost succeeded in stealing the fundamental right of free citizens to establish and transmit their own values by using their enormous resources to propagandize the death of anything sacred in our lives, and to attempt to enthrone empty hedonism and consumption as the ultimate values to guide our conduct.  Free people that we are – and having realized that the ideal model of a market-based economy presented by the corporate apologists is false – we must instead insist on being able to develop and exhibit our own values, and on being able to act on these in all of our economic and political choices, even when these are generated and sustained by institutions other than "the market".

### It's Not the Best of All Possible Worlds

There is a second fatal flaw in the theoretical underpinnings of free market propaganda which is generally known as "Arrow's Impossibility Theorem", after another Nobel Prize-winning economist, Kenneth Arrow[3]. It is generally applied to the analysis of voting rules as means of aggregrating individual preferences into a single "social welfare function" that reflects the preferences of society as a whole, and, basically, shows that there are no known voting rules that lead necessarily to a general result that is consistent with the individual preferences of voters without violating certain basic conditions of reasonableness, such as that, if all voters prefer A to B in the absence of any other candidates, the entrance of C, who is not the first choice of any voter, should not change the result between A and B in favor of B. This condition is known as "independence from irrelevant alternatives", and can be shown to be violated in many both theoretical and practical examples, such as, perhaps, when the participation of H. Ross Perot in the 1992 election may have determined the outcome in favor of Bill Clinton as opposed to the first George Bush.

Arguing backwards, if it is shown by Arrow's theorem that it is impossible to reliably generate a total result from the aggregation of individual preferences by any seemingly reasonable voting rule, it would also seem to be true that it is impossible to assert that any observed total result is necessarily the outcome of an unbiased aggregation of individual preferences. Thus, it would be impossible to sustain that the observed overall result of consumers' activities in the marketplace – a particular set of goods and services produced and consumed – is the only possible or, even less, the "best" aggregation of individual consumers' revealed preferences. Even if it were true, as Samuelson asserts, that individual preferences are revealed by their actual choices in the market, it would still be impossible to say that the overall preferences of society are revealed by the actual outcome of economic activity over any given period.

As though it weren't evident in any case, all of the above is just to say that even in a strictly theoretical sense it is not possible to truthfully claim that the observed outcomes of "free" market activity necessarily reflect the best of all possible outcomes, given the individual preferences of so-called "sovereign"

---

[3] Prof. Arrow's Nobel Prize was awarded for other work delineating the necessary conditions for the existence of Walrasian general equilibrium, though he is probably best known for his Impossibility Theorem.

consumers. "Consumer sovereignty" is bunk, and serves mostly as a diversion used by the corporate apologists as a political tool to lull the voter and consumer into thinking that the manipulation he experiences in the marketplace daily is really consistent with his own and with the general good.

To give an example, although we all see poverty, hunger and disease all around us in the world, very few would want to argue that this economic result is the consequence of consumers making free choices in a free market economy. It would be almost impossible to find anyone who would admit to preferring a world with these afflictions over the alternative, and yet they are a very real and palpable result, especially to those who suffer them.

Conversely, we cannot then sit quietly by and accept assertions to the effect that the good things produced by a free market economy – jobs, income and a reasonable standard of living for a great many people – are simply and uniquely the result of consumers making free choices in a free market economy. If we are to accept the good outcomes of the workings of the system, we must also accept that there are also bad outcomes and that measures can and must be taken to correct them. Those who – having recourse to spurious historical arguments – say that poverty, hunger and disease are inevitable and will always be with us are simply wrong and lead us down the path of irresponsibility and hedonism.

*The truth is that these evils and others that will be briefly analyzed below continue to exist because the system is failing to provide us with the proper choices, and because we are not being taught the proper values. That is the challenge of the current generation: to change the system so as to preserve the good that it produces, while generating the right technical and moral options so that we as humans inhabiting this marvelous planet can proceed to make life here the productive, joyous and humane experience that it is intended to be.*

There is another deep failure of the market system – one that is not even denied in theory – which we must also emphasize. Obviously, the impact of individual preferences and choices on the outcomes produced by the free market economy depend heavily on the relative wealth and power of the individuals making these choices.  Just as obviously, the amount of wealth and power each of us enjoys has very little to do with anything other than the wealth, power or talents that we each just happened to be born with.  Who were our parents, and what did they teach us?  How positive or negative is my outlook on life?  How courageous am I, how "intelligent", how disciplined?  How opportunistic and selfish, perhaps?

All of these things determine our preferences and our ability to act upon our preferences, "freely" or otherwise.  And, although economic theory is content to shrug it off and once again to accept the "inevitable", I for one do not see any reason why Donald Trump's desires to have a personal yacht with golden faucets should "trump" (sorry) my impoverished mother's ability to obtain treatment for her cancer.  There are obviously many other examples of such unfairness in the current workings of the free market system, too many to belabor here.  The point is to be made painfully aware that they do exist, and to forcefully deny their inevitability.

There is no reason to expect or to wish that the free market result in exactly equal outcomes for all people.  There is even something positive to be said about a system that "rewards" things like talent and hard work, even if they are mainly the result of inherited traits and not any particular merit of the individual.  What is not acceptable, however, is for there to be lavish overconsumption and waste of common resources by some, while there are still so many among us whose most basic human needs have not been met, despite, in so many cases, their obvious talents and desperately hard work.

## The Generation and Regeneration of Poverty

While it may be relatively easy to understand how imperfections in the system generate a dynamic which benefits certain powerful constituents when waste products are allowed to be simply dumped into the environment, or when violence or the threat of violence is used illicitly to extract rents or take away the property of others, it is more difficult to understand and especially to accept how the system, as part of its intrinsic functioning, generates and then regenerates poverty on a massive scale.

*The Origins of Poverty?*

To begin with, recall that the "system", though a single integrated structure, is not everywhere homogeneous. Rather, it is composed of a vast number of semi-autonomous constituents – individuals and organizations – that operate a very large number of "sub-systems", some larger and more powerful than others. Like the "system" as a whole, "sub-systems" also exhibit a dynamic whereby their resources and energies are directed toward their own preservation, reproduction, expansion and self-perpetuation.

What happens when some part of the overall "system" is impacted by an external catastrophic event, be it a drought or a hurricane, an epidemic or a Great Crusade? Well, quite simply, the stronger and more powerful constituents included in the larger sub-systems will be able to survive the onslaught in better shape than the weaker among them.

It may long ago have been that all constituents were more or less equal, and it may have been purely random as to who was affected more and who was affected less, but, then as now, almost inevitably some constituents will be weakened and "impoverished", while others will in some way benefit from any catastrophic occurrence.

Some may have simply by chance been closest to the deepest well when the drought occurred, built a fort around it to ensure their own survival, and subsequently became the rulers of an oasis where all who later came to drink had to pay and where subsequently a thriving commerce flourished, built in part with the credits extended by the new rulers from their accumulating surpluses.

Others were not able to find sufficient water, lost their animals and younger children, and were reduced to begging in their old age in order to survive. The surviving older children were not able to continue farming their own land and sold themselves into bondage as field hands and house servants for the more fortunate.

When Katrina hit, the wealthy, whose homes were better built and more suitably located to begin with, were able to get out of the way with plenty of time to spare.  Whatever damage was endured by them was rapidly repaired with the proceeds of their insurance.  The poor, already disadvantaged before the storm, died in large numbers, lost most of their worldly possessions and are still trying to figure out what happened and what will become of them now.

When economic recessions hit, the system allows that corporations and other employers dismiss their workers *en masse*, transforming them instantaneously from self-supporting, productive members of society into wards of the unemployment insurance system, while it lasts, and to candidates for poverty if, for whatever reason (ageism, sexism, racism, etc., or the continuation of recession for too long) they are not able to find another replacement job in a short period of time.

Maybe, in the meantime, the job will have been moved overseas, where the capitalist, who has been able to maintain control over his assets despite the recession by transferring its main impacts onto his former labor force, has found a cheaper and less-demanding source of labor for his relocated factory.

Poverty, it can be said, is in fact the financial and economic safety net of the rich.  It is a mechanism whereby the impacts of external calamity are diverted – naturally and by the un-self-conscious and "innocent" application by the rich of their resources and power to protect themselves and their families – away from the most powerful in society and piled onto those who are less able to protect themselves.

*Poverty as a Consequence of Deliberate Action*

In history and still today, poverty is also often the result of the deliberate actions of other human beings.  As we all know, it was just 150 years ago in the United States that it was still legal and customary to buy slaves who had been captured like wild animals in their African homelands, bound and shipped like cattle in the deadly stinking holds of slave ships, stripped naked, exhibited and sold like hogs, only to be whipped and worked like mules until their eventual deaths, without hope for themselves or their children of rejoining their ancestors in either this world or the next.  It took 100 years after the emancipation of American slaves for a majority of their descendants to achieve anything like civil rights equal to those of other Americans, and a great many still languish in poverty, insecurity and ignorance in the so-called inner city "ghettoes" that are so much a part of the urban American scene.

To this day, throughout the Americas racially indigenous peoples are almost everywhere still suffering the onslaught of a European invasion that began over 500 years ago and spread by savagely unequal wars of conquest that have yet to be ended definitively in some parts of Central and South America. The so-called Indians of the Americas, from the Dakotas to Amazonia and the Andes, are almost all still poor, still undereducated, still culturally repressed and still politically disenfranchised – 516 years since the first landing of Columbus on the American continent.

Once generated, whether by sheer chance and the natural instincts of all people to think first of themselves in times of calamity, or as a consequence of the deliberate actions of rapacious human predators, poverty tends to regenerate itself, generation after generation. The children of teenage crack whores, born into desolation, dehumanized and degraded by their pimps and pushers, will themselves more than likely become crack whores, pimps and pushers. The barefoot child raised in the gutters of a Brazilian *favela* instead of in a school, will more than likely live out his life just like his parents, on the most meager margin of existence, and will more than likely produce children who will also be trapped by poverty and ignorance, even from before their own birth.

It's all part of the system, perpetuating itself until it finally breaks down catastrophically. Or, much more difficult but perhaps still possible, until enough people decide to look reality in the eye and find the courage to act deliberately and probably at considerable sacrifice to themselves, to break the cycle once and for all.

### *The International Propagation of Poverty*

At the international level, poverty is also generated and regenerated by the autonomic functioning of the system. Or, to put it in Biblical-sounding terms, the rich beget rich and the poor beget poor.

23

More crudely, "shit flows downhill", and then tends to stays there.

It is all too easy to forget for those of us who currently inhabit the more prosperous regions of the globe that all, or virtually all, countries that today would be classified as being poor were once the colonies of the countries that today are classified as being rich. The impact of colonization, a process that began in the 15th century and lasted until the 20th, was in most instances devastating to the development of the colonized areas, and has left structural distortions in their economies which persist until this day.

The basic colonial system consisted in the armed conquest of non-Christian, non-European territories, the subjugation of their native peoples and the elimination of their leaders, and their conversion into low-cost producers of mineral or other primary commodities for the benefit of the home country. The amount of mineral wealth extracted from Africa and the Americas, produced by native labor working in conditions of slavery or its equivalent, is truly staggering. Directly or indirectly, this massive transfer of wealth permeated all of the western European countries and can still be seen today in the broad boulevards and parks, the palaces and other monumental architecture that grace many of their cities. Gold, silver, diamonds, copper, bauxite, the finest timbers and many other materials extracted by the millions of tons at the point of a sword or the end of a lash, stolen – for there really is no other word for it – and transferred to the home country to finance the construction not just of palaces and fine buildings, but also schools and universities, railroads and factories.

The former colonies not only suffered from the pillage of their wealth, but also precisely from an almost total lack of investment in any industry or infrastructure not directly related to the extractive process, or in enough schools or universities to allow their populations to develop their potential. The only people who received any semblance of a full education, even by the standards of the times, were the sons of the transplanted colonialists and a few of their administrators who managed the colonies and the extraction of their wealth for the benefit of the home countries.

Thus, when independence was finally achieved, as late as the 1960s in the case of most of Africa, the former colonies found themselves mostly still being governed by local exporter elites whose cultural and economic ties and loyalties remained solidly with the former home countries. They were trapped in economies with only the most meager of infrastructures and that had never developed beyond the extraction and export of a small number of primary commodities, with minerals still being important in some cases, and in the rest consisting of agricultural cash (non-nourishing) crops like rubber, coffee, sugar and tea. Their populations were illiterate in their majority, destitute and riddled by infectious diseases that elsewhere in the world had long been eradicated.

What else to do, when ultimately abandoned to their "independence" but to continue producing the coffee or the rubber that they had always produced, and to continue selling these for whatever price they could receive in the former home countries, if nothing else so as to be able to finance the imports of the local elites, and even if nothing better than bare subsistence resulted for the poor majorities? There was nothing else to do, nor anyone to finance it, perhaps until the late twentieth century produced some effort by the former colonists and the handful of multilateral organizations they created to bring at least a minimum of social and economic development to the countries of the "Third World".

### *A Modern Economic History of Poverty*

In the two decades following the close of WWII, several "multilateral" development banks were created by the victorious former allies to channel resources into the poorest regions of the world. While altruistic motives were undoubtedly to be found among some of their initial organizers, the "First World" countries who capitalized these new institutions were also motivated by proliferating Cold War challenges in the Third World, the growing strength of the "non-aligned" movement and the almost universal repudiation of multinational companies and the role they had assumed in the extraction of minerals and the production of agricultural export commodities in the former European colonies.

The technocrats put in charge of the management of these multilateral banks at first believed that rapid economic development and the eradication of poverty could be achieved by "industrialization" and the creation of basic infrastructure that would allow these economies to become less dependent on the first world for the manufactured goods they had always had to import. By increasing labor productivity through the infusion of capital, these investments would also generate higher paying jobs that would eventually provide a decent modern standard of living for the majority of their populations. In time, it was hoped that the newly-industrialized countries of the world could become competitive on an international scale, and compete successfully with the first world manufacturing exporters.

The idea involved the concept of "infant industries" which would be protected in their home markets by high external tariffs that would keep out first world imports and allow the new third world manufacturers to learn to be efficient.

With the possible and limited exception Taiwan, Korea and the Southeast Asian "tigers", the model was an abject failure, for two interrelated reasons, principally.

First, it was not recognized how the infusion of external capital on concessional terms would create vast opportunities for corruption in states that, besides being poor economically, were almost totally bereft of democratic institutions or the checks and balances that underpin the effective rule of law in more advanced societies. The local elites who one way or another controlled local governments – or the state itself – became the owners of the new ports, power plants and factories, and they – protected from international competition by high external tariffs and a strongly enabling environment both locally and in the international arena –  proceeded to enrich themselves beyond their wildest prior imaginings. Infant industry excuses for the continuation of high tariffs were held to long beyond the period when local industry should have achieved competitive efficiency, and the poor local populations, who before could only import limited quantities of expensive first world manufactures, now could only consume even more limited quantities of exorbitantly expensive, low quality local manufactures. That is to say, in real terms the poor of the Third World only became poorer.

Extended tariff protection and the small size of the domestic market (especially in terms of purchasing power) made it impossible for the new industries financed by the multilateral development banks to grow beyond a very small scale, and therefore impossible for these industries to provide high-paying employment on anything like the scale that would have been needed to make any appreciable dent in poverty and living standards.

But corruption was widespread and strong, and corrupt local elites continued to collaborate with international corporations still primarily interested in the extraction of primary commodities at minimum cost, and the system has endured far longer than one might have thought possible.

The second great failing of the multilateral development banks in the fifties, sixties and seventies was their failure to understand the importance of human capital and investments in education and training in making possible the achievement of rapid economic development and the establishment of minimally functioning democratic institutions.  Except in parts of Asia, where education was and had always been a strong cultural value, very little was done to reduce illiteracy or provide for the training of the technical and managerial cadres that would be needed to operate a modern industrial economy.  Ignorance and illiteracy stripped vast numbers of people of opportunity and the hope of a better future, contributing only to the perpetuation and regeneration of abject poverty in the rural hinterlands and urban slums alike.  Their experience of poverty and hopelessness in these settings has also undoubtedly contributed to a growing poverty of the spirit in forgotten communities around the world, and to the consequent explosion of criminality and antisocial violence that is beginning to victimize us all.

So, accelerated industrialization under the sponsorship of the multilateral development banks in the 50s, 60s and 70s ultimately proved to be a failure.  The late 70s saw the stability of the international financial system threatened almost to the point of collapse by the oil price increases imposed on the world by the Organization of Petroleum Exporting Countries (OPEC). Oil-dependent third world countries saw their oil import bills mushroom and their external indebtedness grow by leaps and bounds.

In an effort to avoid a collapse in world expenditure on other products and the consequent economic and financial havoc this would wreak on the economies of the first world, major international commercial banks were encouraged by the world's financial authorities to "recycle" the petrodollars being deposited with them by the OPEC countries. Specifically, they were positively encouraged by the world's monetary authorities to seek out new large borrowers among the governments and government agencies of the Third World. For a while it was perhaps naively thought that this would bring large amounts of fresh financial resources to bear on the problems of development, and at low risk to the lenders because of the *sovereign* guarantees of the borrowing governments. It was *unheard of* (or so the eager bankers and their regulators pretended) for a sovereign government to default on an international obligation. At worst, if they became overextended, they might have to increase taxes on their populations, but default, never!

Well, things turned out quite differently. Greed spread like a contagion from overeager lenders to unscrupulous borrowers who stole by the tens and hundreds of millions, and pretty soon the poor oil-importing countries of much of the Third World found themselves with levels of external indebtedness that they could never hope to repay, nor even service. In light of payment interruptions, new credits were frozen and a tortuous process began – designed primarily to save the Western financial system and its major commercial banks from collapse – which consisted in the imposition, by the International Monetary Fund, of strict programs of "adjustment", involving almost without exception a drastic devaluation of the debtor-country's currency – presumably to help restore international export competitiveness – and a sharp reduction in government spending other than for debt service. Also common was the forced elimination of any restriction by the debtor government on the international movement of capital, presumably to assure that international debts owed by their domestic private sectors would also be paid. Along with these adjustment programs came the negotiated rescheduling of payments due to the commercial banks and the gradual reduction in principal owed to these banks through the extension of new, very long-term loans by the "multilateral" banks.

Devaluations of course imposed immediate further impoverishment on the already poor countries – drastically reducing the real value of domestic wages in terms of import goods – and fiscal austerity meant further deferral of much needed investments in such things as education, health and basic infrastructure.  The poor continued to get poorer, while the rich – and who can blame them for wanting to do it – managed to hold onto and preserve the value of their financial assets.

The 1980s are often referred to in the trade as "the lost development decade", and, along with financial turmoil on a global scale, were also characterized by armed insurrection and revolution in places as distant and diverse as Peru, Colombia, Nicaragua, El Salvador, Sri Lanka and Afghanistan.  Concerned about the leverage being gained against them by the Soviet Union as a consequence of the financial hardships being imposed on poor countries in the Third World, the United States and certain European countries began to develop programs granting preferential access to their home markets to countries in highly threatened regions, such as Central America and the Caribbean.

The United States, for example, established a program called the "Caribbean Basin Initiative" which set quotas on U.S. imports of certain items such as apparel from low-cost Asian producers, and made these niches available to "friendly" countries in the Caribbean Basin.  U.S. contractors, representing major brands such as Hanes and the Gap, went to the eligible countries of the region, helped to organize local production and supplied raw materials and in some cases equipment to get things started.  Many new jobs, albeit very low paying, were created very quickly and at relatively low cost in terms of investment.  The CBI helped to keep the wolf at bay in these highly unstable regions, as did similar programs developed by former European colonists that provided preferential access to their home markets for the exports of their former African colonies.

But relatively little progress was made in terms of alleviating poverty.  Massive out-migrations began, and the remittances of legal and illegal migrants to the U.S. and Europe have become increasingly important components of poor countries' foreign exchange receipts, in some cases equaling or even exceeding the value of their exports.  So, the poor have now had to invade the rich countries in order to make a living for themselves and their families, but this – despite the evident dependence of rich countries on cheap, imported labor – is beginning to create strong political resistance, nowhere symbolized as dramatically as by U.S. efforts currently underway to build a more than 2,000 mile 12-foot high fence along the whole of the U.S.-Mexico border.

But for the palliatives of programs like the CBI and the safety valve provided by legal and illegal out-migration, conditions in poor countries of the world would surely have reached a crisis point.  The system – in this case the international trade and monetary system – has evolved in such a way as to create a poverty trap that is virtually inescapable.

*The Poverty Trap:  Competitive Currency Devaluations*

In simple terms, what has been observed to happen in repetitive cycles is more or less as follows.

A poor country in crisis accepts the need for external and internal "adjustment", sharply devalues its currency, restricts its foreign borrowing and cuts government spending to the bone. It's "competitiveness" thereby having been restored (i.e. the dollar value of domestic wages having been reduced), assembly industries are attracted and large sewing enclaves are established near all the ports, making blue jeans and underwear for the international market.

Gradually, the poor country's increasing export earnings give rise to demands for wage increases, however modest, and for the increased importation of certain consumer goods.  Who knows, the country might even want to increase its capacity to generate electricity or to provide potable water to its secondary cities, and will borrow and import in order to do so.

Gradually, the exchange rate begins to appreciate from the very low value that had been set by the adjustment program.  Added to the nominal increase in wages granted in response to improved internal market conditions, the country begins to lose competitiveness in comparison with other poor countries around the world, near and far.  Some of the export assembly business begins to relocate to these other countries, simply in response to the "dictates of the market". Our poor country begins to lose foreign exchange earnings but, in an effort to maintain imports of both capital and consumer goods – and thereby protect its hard-won slightly improved standard-of-living – begins to borrow more, albeit at short-term and high rates. Deficit spending also increases in an effort to maintain services, but this only leads to inflation and further pressures on wages and external competitiveness. Debt service begins to pinch until, one day, new sources of short-term credit to repay an ever-increasing quantity of maturing short-term debt suddenly dry up, and the country is once again in "crisis".

What else to do when "crisis" comes? Call out the military one more time to quell the rioting in the streets and accept one more round of internal and external "adjustment" and the further impoverishment that this implies.

This cycle of competitive currency devaluations has been observed over and over again since the 1980s. Sparked the first time by a crisis in Mexico (the "Tequila" round), later by payments difficulties in Russia and then among the southeast Asian ex-tigers – especially Thailand and Indonesia – these cycles have forced Third World countries as a group to keep their currencies and wages cheap in terms of US dollars and Euros, as part of the natural functioning of the system. US and European consumers have thus been kept relatively "rich" in terms of their ability to consume cheap blue jeans and electronics, while Third World populations have been kept "poor" in terms of their inability to consume the barest necessities, including medicines, housing and educational services. Where the distress of poverty has been exacerbated by drought or warfare, famine has continued to be the inescapable result, as in the Sudan.

There is no necessary presence of any malice in the system for it to produce – and continually reproduce – such distressing and deeply undesirable results. It is simply a question of natural self-interest, starting conditions of extreme inequality in wealth and power, and market forces unfettered by any countervailing governmental actions that results internationally – as is evident to any that will open their eyes and look around the world today – in the rich getting obscenely richer and the poor getting poorer

31

and poorer. How long this can continue is a question I, for one, would not like to see answered.

As a closing note on the question of the systemic generation and regeneration of poverty on an international scale, this section will conclude with an analysis of what happens when malice is deliberately introduced into the system, as in the critically-destructive impact that the criminal (there is no other word for it) relationship between the Chinese totalitarian state and large Western corporations is currently having on the poor all over the world, including the poor that live inside the rich First World economies.

## Gangsta' Trade – China's Economic Relations with the West

In the early 1970s Richard Nixon and Henry Kissinger launched a new policy of "Entente" with the totalitarian regime governing the People's Republic of China. This gradually led to diplomatic representation, membership in the UN and other international organizations, diplomatic recognition by the U.S. and, despite the violent repression of civil liberties by the Chinese government in 1989, to growing commercial relations between China and the West. The U.S. and other Western interest in developing strong commercial relations with China doubtless stemmed in part from the belief that bringing China into the international trade system would help to bring about political change in China, and reduce the chances of armed conflict. It also undoubtedly stemmed from very strong profit motives by Western corporations, who saw in China both a huge potential market for Western capital and consumer goods, and a vast pool of cheap labor available for the production of consumer goods for the international market. In 2001, China, almost as if to a religious cult, "acceded" to the World Trade Organization and became eligible for "most-favored nation" treatment under the WTO's Generalized System of Preferences, or GSP.

This might all be well and good but for the fact that, unlike any other Third World member of the WTO, China never has and still does not play by the rules. Their huge economic power – tightly controlled by the state political apparatus – has made possible the development of huge business undertakings by Western companies selling into China or, especially, buying/sourcing ridiculously cheap consumer goods in China, which they then market in their own home markets at great rates of profit. The vast sums of money being made by Western companies that have organized this trade have been judiciously applied, in part, to the management of public opinion regarding trade with China, and to the election and re-election of government officials who can be relied upon to continue supporting the *status quo* insofar as the terms-of-trade with China are concerned, despite the terrific harm that is being done to Western and other competing Third World industries, and especially to the labor forces in these regions who are unable to compete against the special advantages that China is being allowed in its dealings with the West and – what else? – are losing their jobs by the millions as a result.

What are these special advantages?

First, as has been widely commented upon but which has to be seen to be believed, China violates labor and environmental norms to an extent that would cause international uproar anywhere else in the world. Public resistance to such violations within China is repressed by what continues to be a totalitarian police state, information about what is going on is suppressed, and the Western business partners benefiting from these abuses are relied upon to manage reporting and public opinion in their own countries.

Chinese industrial wages, even in local currency terms, are meager in comparison with other Third World countries, and workers are treated almost as indentured servants, obliged in many cases to be separated from their families and to live in factory barracks where few amenities are available. The use of child labor and that of prisoners in industrial facilities producing for export to the West has been widely reported. The health of the population is damaged by uncontrolled pollution of air and water with the toxic emissions of industry, and the destruction of the Chinese environment imposes costs on current and future generations, in China and the rest of the world alike.

All of these violations of minimum norms observed in other parts of the world serve to reduce the costs of Chinese goods, helping to drive competitors from the market.

Just as egregiously and harmfully, with the apparent complicity and connivance of the world's monetary authorities, the WTO and the leading governments of the Western nations, China has been allowed to maintain a hugely unfair competitive advantage in international trade through the maintenance of a grossly undervalued currency exchange rate.

Organizations such as the World Bank and others who have conducted studies based on purchasing-power parity have estimated that the Chinese currency is valued at only about 40% of what it would and should be worth if unrestricted exchange market conditions were allowed to prevail in China by its government.  This means, simply, that Chinese goods should cost about *2 ½ times as much* in US dollars or Euros as they currently do[4].  This is obviously a huge competitive advantage in the international marketplace, one that has allowed Chinese exports to displace formerly competitive manufactured exports from other Third World countries to the US and Europe, and which has also significantly damaged the manufacturing industries of those First World regions themselves.  The occasional 2 or 3 percentage point exchange rate adjustments that the Chinese government has allowed in recent years in response to blustering Western parliaments and press are nothing less than a mockery of the much-vaunted international trading system represented by the WTO and its key members.

Still further, the Chinese government continues to subsidize Chinese industries when it considers it to be strategically necessary.  How?  Primarily by failing to collect on the loans extended by state-owned banks to state-owned export industries.  When pressed, Chinese officials will make statements to the effect that efforts are being made to reorganize the indebted industries and collect on overdue loans, but time passes and nothing happens.  No one is in a position to make anything happen, nor really wants to, so that, in time, loans become public subsidies and grants, violating every precept of free market economics and fair international exchange.

---

[4] Indeed, the 2007 CIA World Factbook estimates the Purchasing Power Parity exchange rate for China in 2006 to be *4 times* the official dollar exchange rate.

Why would the Chinese government want to absorb the costs imposed on its own economy by these practices?  In two words, and just like the old-time robber barons, to gain *market share* and the market power that comes with it.  Predatory pricing has always been an instrument of would-be monopolists, and is always intended and used to ruin weaker competitors, achieve control over the target market and generate vast quasi-monopolistic profits in the future.  In only about two decades, China has gone from almost zero exports to the United States to a position where China, with an approximately 16% share, is about to overtake Canada as the largest single source of U.S. imports.  Their share of U.S. imports is already more than double that of either Germany or Japan, erstwhile among our largest external suppliers.  Similarly large market shares are being developed in other international markets in Europe and even in the Third World, and China has become – along with Japan, Taiwan and Korea – one of the most important external creditors to the U.S. government, holding a substantial portion (about 15%) of the over $10 trillion in U.S. external debt.

China has become indispensable, both as a source of low-cost consumer goods to help maintain our so-called high standard of living, and as a creditor on a massive scale that allows us to keep the party going.  We are addicted and we are caught.  We will rue the day when – as almost inevitably will happen if things continue as they are –the Chinese government decides to begin tightening the screws they have so laboriously put into place, and sharply increase their exchange rate and the export value of their manufactures, while simultaneously refusing to continue rolling over maturing U.S. government debt.

How does the Chinese government finance the costs of these policies and practices?  As a centrally-managed totalitarian government with both the means and the will to carry out its agendas, the Chinese government finances the cost of destroying its international competition – including perhaps especially any competition that remains within its principal target markets – simply by forcing these on its own population, especially the less-favored rural sector.  Buying market share and destroying international competition essentially means subsidizing American and other international consumers, at least temporarily, while imposing a much lower real standard of living on its own population than would otherwise be the case.

There is no "free market" in foreign exchange in the People's Republic of China. The government rations FX to its favored industries to enable them to import capital goods and raw materials, while the remainder of their huge current account surpluses (currently about $250 billion annually on export sales of about $1 trillion) is *sterilized* in the form of investments in foreign financial assets, especially U.S. Treasury obligations. Thus, a large source of potentially increased domestic demand, including for imported consumer goods, is choked off at the source before it can be translated into increased domestic prices and wages. Further, as was mentioned above, labor conditions in China are such that there is no question of Chinese workers organizing or striking for better pay and working conditions. Their costs, even in domestic currency terms, are kept extremely low – in part by tight government controls over access to jobs and internal labor mobility – and in part by the overhang of a huge underemployed and deeply impoverished rural population that, like Marx's "reserve army of the unemployed" stands ready at any time to provide the marginal replacement worker at a pathetically low nominal wage. The Chinese labor force is currently estimated at about 800 million men and women, of which 45% or about *360 million people of working age* are in the underproductive and underemployed agricultural sector. In addition, it is estimated that there are already between 100-150 million *surplus* rural workers, who, according to the CIA, *are adrift between the villages and the cities, many subsisting through part-time, low-paying jobs*.

So, the Chinese government is able to carry out its aggressive international economic programs because it is strong enough to be able to force the costs on its own people. When the next stage of their program – converting their power into the ruin of the leading Western economies and achieving (recapturing, in their view) long-term hegemony – is ready to be implemented in the not very distant future, they will be able to do so by once again forcing a domestic belt-tightening – something Western governments are basically unable to do in any circumstances short of depression and/or war – and by drawing on the *trillions* of dollars in international reserves that they are rapidly accumulating.

Why is this blatant and destructive abuse of economic power and the international trading system allowed to continue by the rest of the world?  The answer here is simple, too.  It is allowed to continue because there are some very large and very powerful corporations in the U.S. and elsewhere in the West that are making huge amounts of profit from their ownership of brands of mass-market consumer goods that – manufactured at artificially low cost in China – earn huge margins when sold into their domestic markets.  These corporations, whose self-interest has reached an extreme even by Smithies standards, ensure that the Chinese fix continues in place by using a small portion of their profits to line the political campaign chests of politicians on every side of every aisle.  These, of course, are easily convinced by all these dollars of the economic wisdom lying behind the *laisser-faire* position, and the fix continues despite the blatant destruction it is causing to the Western industrial economy as a whole.

The fix, please recall, consists of the combination of the following manipulative elements that have been universally outlawed, except in the case of the Chinese and their marketing allies:
- grossly substandard labor and environmental norms
- a managed, and egregiously undervalued, foreign exchange rate
- a restricted foreign exchange market
- a controlled external market, on both the import and export sides
- a restricted internal labor market
- direct subsidies to strategic export industries, disguised as credits
- political dictatorship and repression

For all of the chatter one hears on television about the harm being done to the U.S. economy by NAFTA and by CAFTA, the reality is that none of the above-mentioned abusive practices are allowed of signatories to these two trade agreements.  The problem, Mr. Lou Dobbs, is *CHINA*, not NAFTA or CAFTA, and the problem continues because our own economic and political leadership has sold out to the enemy.

## The Destruction of Our Environment

In the previous chapter, we showed how the "system" works –
along with all the other good things it may also do – to generate
and then to continually regenerate poverty. Poverty is not an
accident. Poverty is not an omission – a certain group of people
having been inadvertently left out and left behind. Poverty is
the direct and simple consequence of chance, unequal power,
and values – which are also a product of the system – that do
not encompass the need to remedy the negative impacts of
social interaction. We showed how, without any necessary
malice or conspiracy on anybody's part, poverty could be
generated within a society of equals simply by the something as
arbitrary and unpredictable as a drought. Poverty is then
regenerated by the simple pursuit of self-interest by the
members of society, who, because of the random impacts
suffered by the poor in the first round, now have unequal power
and unequal ability to protect themselves. The simple, naive
and even, "innocent" pursuit of self-preservation and self-
replication by the "rich", has negative consequences for the
"poor", making the regeneration of poverty an intrinsic feature
of a system that does not incorporate a "countervailing power"
mechanism to redress the wrongs that may initially be the result
of nothing more than chance.

Having understood the mechanism whereby poverty is
generated and regenerated, we then explored the workings of
the system in the modern world we live in, and found numerous
examples of ways in which the unfettered pursuit of the
interests of the powerful few, leads to the perpetuation of
poverty among the many. Public institutions are corrupted and
co-opted by the powerful – or stifled before they can develop –
so that they cease to operate correctively and are
reprogrammed to serve mainly to justify and perpetuate the
current state of affairs.

Shit flows downhill, and then stays there.

Gold, however, seems to float. In the first instance, because
nothing is done to prevent it. In the second, because most of
the resources of society are directed, by those who control
them, to keeping the pumps in good working order. The
system, thus crudely but accurately described, functions in this

manner just as much within the borders of the countries of the first world, as in determining the outcome of cross-border relations between the rich and poor countries of the world.

How does the system also contribute to the destruction of the world's natural resources? This, as indicated at the outset, is the second great evil facing the world today, and if we are to keep it from foreclosing on the futures of our grandchildren, we must understand how it too represents – not an accident or an omission – but just another facet of an imperfect economic and political system.

World fisheries are down to only about 10 percent of the wild stocks that existed in our oceans and seas only 50 years ago. As put in a recent interview by Daniel Pauly, Professor and Director of the Fisheries Centre at the University of British Columbia:

> Now, let's not forget it, most of the technology we throw at fish is military technology. All the acoustic equipment was developed, for example, in WWII by the allies in chasing German submarines. The GPS, which is the global positioning system developed during the Cold War to position things and to study the Earth in great detail, and now this technology is available to everybody to catch the last fish as if it were, I don't know, a Soviet tank.
>
> *So it's in fact a war on fish you're describing.*
>
> Yes, it certainly is a war on fish, and what I'm saying is that we're winning it; we have won the war on fish.
>
> *We're going to wipe them out to the very last one.*

Add to the impact of military technology on the industry's ability to harvest fish, the further impacts of proliferating unregulated pirate fishing fleets and their use of dynamite and cyanide to stun and capture fish, and the effects of mangrove destruction, waste, fertilizer and pesticide runoff in poisoning the habitat and the flesh of many marine species, and you can see why a growing number of marine biologists are saying that we should no longer eat fish in any quantity, despite its well-known nutritional benefits for mankind.

Fertile topsoil is being lost at an alarming rate, estimated at about 20 percent of the world's soil resources between 1950 and 1990.

Despite massive publicity and rhetoric, forests continue to be destroyed around the world, especially in the tropics but also in the great Siberian plain, to create space for both subsistence and plantation agriculture in the former region, and simply to harvest slow-growing species for lumber and pulp, in the latter. *Annual losses of forest cover* are currently equivalent to the total land area of countries like Greece, Syria, Hungary or Nicaragua.

Finally, experts predict that the world will irretrievably lose about 20 percent of all species currently living on Earth, in just the next 30 years. About 23 percent of mammals, and untold numbers of plants, insects and fishes, will be gone forever.

By what right and under whose authority is such massive and irretrievable damage being done to the world's resources? How can we continue to allow such wanton and irresponsible destruction of our children's birthright? What used to be ignorance has now turned into crime, and whatever money is being made by the industries that are responsible for this pillage is nothing more than theft. It needs to stop. They need to be stopped by governments led by politicians who have a sense of responsibility and can somehow find the courage to say no to the millions that are proffered to them by the perpetrators.

Why does industry persist in production practices that destroy the very basis of their own activities? Simply because there is a lack of enforcement of rules that would force them to absorb the costs that they are now externalizing and imposing on society at large, including future generations, and because there is no mechanism in place to identify and punish "free riders" who are difficult to exclude from the industry in question, despite their abusive practices.

As an example, despite the gross overfishing of most of the world's fisheries and the long-term damage this is causing – including the possible extinction of numerous species – not only is the world's fishing industry not being made to absorb the costs they are imposing on the rest of us, it actually receives massive subsidies – estimated at as much as 30-40% of the total cost of their catch – from most countries having a large fishing fleet.  Furthermore, so-called pirate fleets operating under a variety of flags of convenience, continue to fish in protected waters and to utilize prohibited technologies, despite any good-faith efforts the legitimate fishing industry may be making to reduce the harm they are doing.  So governments – far from protecting the long-term interests of its citizens – are actually subsidizing the continued accelerated destruction of a common resource, while failing to devote the resources necessary to control the abuses of self-proclaimed pirates, out only for the quick buck despite the irreversible destruction of common resources that they may be causing.

Though some would have it so, that's not what I call democracy in action.  That's not the way the system ought to work, even if that is the way things have always been.

The story is pretty similar in the case of other resources, ranging from the management or mismanagement of topsoil, potable water and forests.  Lax and corrupt governments continue to allow private interests to slough off a major portion of the costs stemming from their "productive" activities onto other segments of society, and even take hard-won tax revenues paid in by citizens to subsidize overexploitation across the board.

Why does it happen?  Because we let it happen.  Until we, as well-informed and responsible citizens of Earth demand that these abuses be halted, you can be sure that they will continue. That is just the way the system works.

Alienation, Criminality, and Violence

How, you may ask, can such obvious and extreme evils be construed to "part of the system"?

In significant measure, I would answer, these evils are derived from the extreme poverty of large numbers of people that – as we saw earlier – is perhaps randomly generated at the outset, but which is then systematically regenerated as part of the natural working of the system.  Along with poverty clearly come insecurity, fear, hopelessness, anger and resentment and the alienation that can easily lead, as in the case of our inner-city gangs, to criminality and violence. These, which for further systemic reasons yet to be enumerated, provide a livelihood – perhaps the best that is available – to large numbers of poor youth, become part of the culture and the "juice" that is regenerated and transmitted from generation to generation within the confines of these ever-larger islands of poverty proliferating everywhere in our midst.

Beyond that, I would argue, alienation exists, is produced and reproduced by the system at least in some measure because alienated and disorganized individuals are much easier to manipulate and "govern" – much easier to control – than are people joined together by common values and beliefs in independent organizations. The extended family is all but dead in much of the so-called First World, and even the nuclear family is disintegrating. Divorces now outnumber marriages, and, for economic and other reasons, children tend to leave the home immediately following high school and scatter widely so that further family interaction is reduced mainly to occasional reunions for a holiday or other special occasion. The old, in this so-called First World, are ever more frequently institutionalized, by their families when they can afford it, otherwise by the state. In ever increasing numbers, they are simply abandoned and left to end their days in the street.

Having few family or community ties, few if any strong and reliable social supports, individuals in this brave new world are raised on a constant din of violent and alienating messages – ranging from a fundamental worldview based on the claim that Darwinian tooth-and-nail savagery and competition are hard-wired within us – to the constant, imminent threat of "terrorism", to the distrust and fear of teacher, pastor and neighbour. Alienated individuals living in this manner ultimately lose the ability to act, or even to think for themselves. With few human interactions or alliances outside of work and the immediate family, they become creatures only of their work, the television set, the shopping mall and the sports bar, and constitute virtually no threat or challenge to the established order of things and those who profit by it. Divide and conquer. Isolate and dominate. It's very much the same idea.

42

Crime and violence may grow partly out poverty and partly out of the deliberate alienation of the individual, as was seen above, but they also exist because they are big money-makers for powerful people and institutions, and the ultimate means of assuring the dominance and perpetuation of the system, respectively.

The retail value of the international drug trade is estimated at something in the neighbourhood of $400 billion dollars annually. That is more or less the entire GNP of Argentina, a country of over 40 million people. How much of that money doesn't get spread around, from the peasants to the processors to the smugglers and the *bandidos* who produce and distribute the drugs, to the cops, judges, mayors, DAs, even ministers and presidents who facilitate the traffic? How anxious are all of these people to see the drug trade disappear? Especially, from the point of view of the controllers of the system, when drugs are also such a convenient and effective way of distracting youth, at once weakening their character and their resolve and lessening their ability to get into other, more political, kinds of trouble, while at the same time providing the authorities with a pretext for incarcerating the more wilful and rebellious among them.

Is it any accident, and is there no political significance to the fact that in the U.S. the vast majority of the approximately 2 million prison population is made up of young black men from impoverished inner city ghettoes who have been jailed on drug-related charges? While the middle-class white youth of America is lulled into the la-la land of drugs and Californication? Big profits and social control, all wrapped up into one.

White-collar crime of all kinds is rampant in the U.S. and other parts of the First World, even to the extent of causing major disruptions to such major components of the system as the stock market, and the real estate and housing finance markets. At its criminal fringes, the system is out of control and skirting collapse. Taking their lead from frauds like Michael Milken and Ivan Boesky who made their unscrupulous millions in the nineteen-eighties, the chief executives and chief financial executives of such once respectable companies as ENRON and Arthur Andersen cheated and stole their way to millions in the nineties, helping to fuel a financial bubble in the stock market that finally burst when their criminal wrongdoing was finally made public. Having lost confidence – the prime asset of a public financial utility like a stock market – the investing public including major institutions managing billions in pension funds and insurance reserves plunged headlong into the real estate and mortgage-backed securities markets, transferring the bubble from one segment of the economy to another and only deferring the pain which is now beginning to be felt by the general public as a result of the excesses of the system.

The following list, copied word-for-word from the Internet pages of Wikipedia (see "accounting scandals"), gives an idea of just how badly the system has gotten out of control in terms of the blurring – or even the obliteration – of the lines between the honest pursuit of profit and out-and-out criminality.

### *List of companies involved in scandals*

### Big Four major audit firms

(**Audit firms** are listed, followed by select clients ensnarled by accounting scandals)

- **Deloitte & Touche**: Adelphia, AES, Duke Energy, El Paso, Merrill Lynch, Reliant Energy, Rite Aid, Parmalat
- **Ernst & Young**: AOL Time Warner, Dollar General, PNC Bank, Cendant, HealthSouth
- **KPMG**: Citigroup, Computer Associates, ImClone, Peregrine, Xerox, Siemens AG, Banco Nacional S.A. (Brazil), BMW Group
- **PricewaterhouseCoopers**: Bristol Myers, HPL, JP Morgan Chase, Kmart, Lucent, MicroStrategy, Network Associates, NKFS, Tyco

### Predecessor and other U.S. audit firms

- **Arthur Andersen**: CMS, Cornell, Dynergy, Enron, Global Crossing, Halliburton, Liberate Technologies, Martha Stewart Living Omnimedia, Merck, Peregrine, Qwest, Sunbeam Products, Waste Management, Inc., WorldCom. Arthur Andersen was a former major audit firm that began to unwind its operations in 2002 after being indicted for

obstruction of justice for shredding documents related to its Enron audit.
- **Coopers & Lybrand LLP**: Network Associates, Phar-Mor. Former major audit firms Coopers & Lybrand and Price Waterhouse merged in 1998 to become PricewaterhouseCoopers (see above).
- **Gutierrez & Co**.: Vivendi
- **Grant Thornton**: Parmalat

Source: Wikipedia

How is it *possible?*

- Merrill Lynch?
- RiteAid?
- AOL Time Warner?
- Citigroup?
- Xerox?
- JP Morgan Chase?
- Kmart?
- Lucent?
- Halliburton?
- Merck?
- Sunbeam Products?
- WorldCom?
- Vivendi?
- Parmalat?

What major corporations *are not* engaged in criminal fraud against the public? What's happened to our system? Who's in control, and where are they taking us?

Beyond crime itself, violence of every kind engulfs the modern world, especially the most *Westernized* parts of it, beginning with the United States. A great deal of the violence we are deluged by originates in criminal activity of one kind or another, and is largely explained by the same factors.

Then there is all the random, antisocial violence that stems from alienation. Young boys killing their parents and siblings in their own homes. Young mothers drowning their babies, or stuffing them into garbage containers. School children stalking the halls with automatic rifles, mowing down their companions and their

45

teachers, then turning their weapons on themselves. Nothing more than suicide acted out and dramatized, really.

And what of the all the lonely suicides? Even the Scandinavian countries – where murder is almost unheard of – are seeing rates of self-destructive violence that could never have been imagined among earlier generations.

Violence of the most extreme kinds is everywhere in films and on television – perverted, gory, *extreme* – transformed into cartoons to make it Ok for children (?!) and *always* present, whether to terrify us into passive submissiveness, egg us on to self-destruction, or numb us into catatonia, I don't know.

*Why* is there so much violence in films and on television? Is it there perhaps to sell us the inevitability of the other forms of institutional violence that pervade our system, to inure us into accepting its existence as simply a fact of life? As suggested above, *some* kinds of violence are related to criminality and, consequently, to the societal factors that generate criminality. *Some* kinds of violence are manifestations of extreme alienation of individuals from one another in our modern societies, and, consequently, to the factors that engender alienation.

But what of organized, calculated *institutional* violence – ranging from political assassination to the waging of war for speculative profit and the control of resources – that has also become part and parcel of the system and how it works? What of the violence that is deliberately and coldly perpetrated by our governments against our own citizens and our neighbours?

It used to be – at least in law if not always in practice – that nations engaged in violence against each other only as a last resort, and only with the clear, informed support of a majority of the populace as expressed by the formal approval of the people's elected representatives. Now, we are happy to launch missiles or engage in bombing raids over populated areas – killing and maiming men, women and children alike – simply on suspicion that *terrorists* or other phantom enemies may be harboured there. Full-scale invasions and wars – costing thousands upon thousands of lives, massive destruction and the wasting of hundreds of billions of dollars – are launched by so-called *democratic* governments without any formal declaration of war by any elected legislature, and simply on the *suspicion* that the erstwhile ally now so-called enemy, *might* – despite much evidence and much testimony to the contrary – be producing weapons-of-mass-destruction.

Even now, after imposing years of needless suffering on our victims and our own selfless troops and their families, our so-called leaders and their supporters talk blithely and openly of extending war into neighboring and distant countries – now Iran, now North Korea – dreaming, I suppose, of the day when we can say that we have once and for all blasted all our enemies to hell, and, last man standing, have brought into being among the corpses and the smoking ash a brave new world order shaped in our very own blessed image.

It's hard to sort out just how many pathologies are involved in all of this, but they are numerous, deep-seated and complex. As is evidenced by the facts, they have taken hold of a dangerously large number of our leaders, and they threaten to overwhelm our culture.

*Live Free, or Die Hard!* What a great title for a movie. Gee, could we see it again, Dubya? Pleease?

Other top 10 features for 2007:

> *No Country for Old Men* – (do they mean the United States?) The story begins when Llewellyn Moss finds a pickup truck surrounded by a sentry of dead men. A load of heroin and two million dollars in cash are still in the back.

> *There Will Be Blood* – A sprawling epic about family, greed, corruption, and the pursuit of the American dream.

> *The Bourne Ultimatum* – All he wanted was to disappear. Instead, Jason Bourne is now hunted by the people who made him what he is. Having lost his memory and the one person he loved, he is undeterred by the barrage of bullets and a new generation of highly-trained killers.

> *Zodiac* – A serial killer walks the streets of San Francisco in the mid 1970s.

Also of note, 2005:

> *Syriana* – A career CIA undercover operative has been
> assigned a mission in Beirut. He is caught up in a secret
> plot against a Persian Gulf Prince. An oil stock broker, who
> has become a friend to the Prince, advises him on how he
> can make his country better with the oil revenues once he
> becomes Emir. The oil companies don't want the Prince to
> become Emir, because he wants the U.S. military bases
> out of his country, and he wants to build the country's
> infrastructure. The Prince also wants to make peace with
> others countries in the Persian Gulf and not waste money
> on unnecessary items such as expensive warplanes. Once
> they are united, they will control their own destiny by
> controlling their own oil. Then we have a merger of two
> U.S. oil companies under review of the Justice Department,
> and the Pakistani teenager who loses his job at the oil field,
> and is recruited to be a suicide bomber. They all get blown
> up in the end, but the oil continues to flow.

Criminality, alienation, the unfettered pursuit of wealth and
hegemony. Art reflecting life and right back at you (all over
again).

# III:  What Happened?  How Did It Happen?

At this point, we may perhaps have satisfactorily shown that – left entirely to its own devices – the free market system, despite being such an important and effective means to mobilize individual effort and produce efficiently and well, cannot be relied upon to resolve the outstanding critical issues of our time: poverty, the destruction of the environment and the disintegration of society.  This is nothing new.  As stated earlier, such stalwarts of democracy and capitalism as Theodore Roosevelt, Franklin Delano Roosevelt, and John Fitzgerald Kennedy all recognized and accepted the fact that, as Galbraith put it, the government – acting in the name of and for the common good – had to be ready to exert a "countervailing power", to offset the power of the "military-industrial complex" (Eisenhower) and guide the functioning of the market-based economy to produce a more equitable, sustainable and just outcome.

What then, is going on?  Why doesn't our government currently even attempt to fulfill this role?  Why is the general message put out by the propagandists and the prevalent conventional wisdom in countries like the United States that government must not interfere in any way with the workings of the free market economy, despite the failures and shortcomings that stare us so blatantly in the face?  Why do we continue to swallow it all?

The answer, as all of us already know, is money and power.  Money, and the structural distortions in our system that money has been able to gradually put into place, today results in something that looks like anything but "consumer sovereignty" or "freedom".  What the system looks like, increasingly in the U.S. and elsewhere in the "First" world, is an undercover, hidden dictatorship of corporate interests who have co-opted our governmental and other public institutions, especially the media, to establish and maintain a stable environment for their variant of productionist materialism, an ideology that holds (1) that there is no God/that God doesn't matter; (2) that the highest good is wealth, achieved by efficient production through advanced technology and work; and, (3) that the more an individual is able to accumulate and to consume in his/her life, the better.

# The Hijacking of our Government and Public Institutions

It is well, perhaps, at this point to draw on the insight of scholars who have come before us and who – more than fifty years ago – have provided a framework of analysis that continues to be valid to this day.

We quote extensively in the section below from a book published in 1956 by the American political scientist and sociologist C. Wright Mills, which he called The Power Elite.

"Authority", he wrote, "is power that is explicit and more or less 'voluntarily' obeyed; manipulation is the 'secret' exercise of power, unknown to those who are influenced. In the model of the classic democratic society, manipulation is not a problem, because formal authority resides in the public itself and in its representatives who are made or broken by the public. In the completely authoritarian society, manipulation is not a problem, because authority is openly identified with the ruling institutions and their agents, who may use authority explicitly and nakedly. They do not, in the extreme case, have to gain or retain power by hiding its exercise.

Manipulation becomes a problem wherever men have power that is concentrated and willful but do not have authority, or when, for any reason, they do not wish to use their power openly. Then the powerful seek to rule without showing their powerfulness. They want to rule, as it were, secretly, without publicized legitimation. It is in this mixed case – as in the intermediate reality of the American today – that manipulation is a prime way of exercising power. Small circles of men are making decisions which they need to have at least authorized by indifferent or recalcitrant people over whom they do not exercise explicit authority. So the small circle tries to manipulate these people into willing acceptance or cheerful support of their decisions or opinions-or at least to the rejection of possible counter-opinions.

Authority formally resides 'in the people,' but power is in fact held by small circles of men. That is why the standard strategy of manipulation is to make it appear that the people, or at least a large group of them, 'really made the decision.' That is why even when the authority is available, men with access to it may still prefer the secret, quieter ways of manipulation.

But are not the people now more educated? Why not emphasize the spread of education rather than the increased effects of the mass media? The answer, in brief, is that mass education, in many respects, has become another mass medium.

The prime task of public education, as it came widely to be understood in this country, was political: to make the citizen more knowledgeable and thus better able to think and to judge of public affairs. In time, the function of education shifted from the political to the economic: to train people for better-paying jobs and thus to get ahead. This is especially true of the high-school movement, which has met the business demands for white-collar skills at the public's expense. In large part education has become merely vocational; insofar as its political task is concerned, in many schools, that has been reduced to a routine training of nationalist loyalties."

Teddy Roosevelt fought against the overt form of authoritarian abuse of our democracy by the robber barons and their trusts. He made some headway for a while, but not enough to keep them from roaring back in the 1920s, when their blatant abuses and excesses led directly to the Crash of 1929 and the Great Depression that lasted for over a decade afterwards. During that time and in World War II, Franklin Delano Roosevelt was able to design a system in which the power of government was at least intended to be exerted in the public good and to some degree as a counterweight to the unfettered power of the private corporate sector.

But something began to happen after World War II. C. Wright Mills could see it in the early 50's and Dwight David Eisenhower warned us about it in 1960. John Fitzgerald Kennedy tried to fight against it, famously taking on the steel industry in one case, and the CIA and the military establishment in another.

He was killed. And not, as the manipulators would still have you believe, by a "crazed lone gunman". A huge body of evidence that has been meticulously gathered over the years indicates very strongly that he was killed by individuals associated with the Mafia, elements of U.S. intelligence involved in undercover efforts to remove Castro, and the anti-Castro Cuban exile community. The murder was covered up by the U.S. government at the highest levels, whether to hide further complicity or to shield the public from an ugly and perhaps destructive truth, is yet to be determined.

But the fact is that he was deliberately killed by powerful groups acting behind the scenes, and that since his death, the corporate sector in the U.S., mimicked to a greater or lesser extent in the rest of the "Western World" – a world defined by the great Cold War struggle against soviet communism – has been continually consolidating and extending its power and control over most aspects of daily life, economic, cultural and political. Other once powerful institutions such as academia and the Church have been systematically weakened, and the military-industrial-government-communications complex is now solidly in control. Despite the brave attempts of a few isolated writers and media personalities, it is likely to remain so into the indefinite future.

Why? Because, over the course of decades, organically and in response to the natural dictates and demands of power, they have been able to restructure our system. It is now *their* system, and it functions to regenerate and extend *their* power, first – while keeping us quiet, second. Unless and until a large number of us realize what has happened, understand at least in part how it has happened, change our own thinking and mobilize in force to reclaim our "sovereignty" over our economic, political and cultural life – that is simply the way things are going to continue to be.

How has it happened?

Some Design Mistakes

Maybe, to begin with and in part, because our Founding Fathers and the 18$^{th}$ century European ideologues of democracy sold us an unrealistic and impractical vision of equality that had to be restricted – albeit surreptitiously and only in the "small print" – from the very outset.

"One man, one vote" has enormous simplistic appeal as an ideological slogan, but it is a very radical notion and has never been allowed to function fully, and probably never will be, until much greater equality of means and of education is brought into being on this earth.

At the outset in the United States, for example, the vote was limited to free, white males who, in addition, met one or another of their individual State's property requirements. Furthermore, actual governance was done indirectly by elected representatives, introducing an additional distance between the voter and his government, and an additional element of conservatism into its manner of governing.

Women were generally not educated to a high degree at that time, and therefore could not be expected to be knowledgeable of world affairs. And how, thought the Founders, could an indentured laborer recently emigrated from the slums of London possibly have the new Nation's broader interests nearly so close to heart as the owner of a plantation or a mill who had risked all in order to see it become free? Aside from the fairness or unfairness of it, how in practical terms could someone without any education or even the ability to read be expected to understand the nuances of good government to the same extent as the classically-educated elites, steeped as they were in world history, from the times of Pericles to their own?

All rhetoric aside, as a practical matter, they obviously thought one-man one-vote could not be expected to produce security or good government, and restrictions were instituted into the system from the outset.

In Europe, France's early experiment with a radical form of democracy was crushed and, with much localized ferment perhaps, one form of monarchy or another – linked to colonial imperialism – continued to hold sway at least until the early part of the twentieth century. England certainly moved far in developing its parliamentary democracy during the 19th century, but always under strong checks exerted through the House of Lords by the aristocracy and, more directly, by the Monarchy itself.

So it should come as no surprise that, when women and other formerly disenfranchised groups were finally able to win a measure of equality, and as other direct conditions restricting access to the vote were gradually removed, the ruling classes – the establishment that controlled the wealth of Western societies – turned to other means to try to ensure that the outcomes of the "democratic" electoral processes that had been forced upon them would continue to meet their fundamental need for security, in their persons, in their material possessions and in the positions they occupied within society – i.e. in their wealth and in their power.

For many decades extending well into the beginning of the 20th century, the governing elites in the U.S. and elsewhere commonly resorted to a wide variety of primitive means to keep certain population groups from exercising their right to vote, to buy the votes of others, to stuff ballot boxes, and in general to "fix" the outcomes of elections held at all levels of government. Similar practices are continued to this day in many poorer countries that call themselves democracies, but where effective power continues to be highly concentrated and electoral practices are not subjected to any kind of serious public scrutiny.

In the more "advanced" democracies, more elaborate and sophisticated means have been developed to retain effective control of the electoral process for the ruling elites.

The Creation and Takeover of a Mass Political Market

Early in the 20th century in the United States, and perhaps somewhat earlier in Western Europe, a process of nationalization and consolidation of the major media began to take place. Newspapers like the *Wall Street Journal* and the *New York Times* began publishing regional editions in various parts of the country, on their way to becoming national journals. Other precursors of today's giant media conglomerates, the Hearst and Pulitzer newspaper chains, began buying up regional papers across the country, many of which maintained their local identities but became subject to centralized editorial and commercial control. These developments, later replicated in the radio and television industries, made it possible for the first time to generate and disseminate a unified national message, addressing – indeed in part helping to create for the first time in history – a unified national market for everything from corn flakes to political ideas and political candidates.

Access to the national market – as well as more narrowly defined state and local markets – became indispensable for any candidate seeking national office.  But access to the national market doesn't come cheap.

Media, and more specifically television, have become the gateway to political office in the modern world.  Without paid access and advertising expenditures that collectively now run into the billions of dollars in a presidential campaign in the US, no candidate can even become known by, much less capture the imagination of, the voting public.  Political candidacy and political office, especially at the national level, is now the restricted domain of either the very rich or the very beholden to the very rich, in particular the major – now *trans*national corporations and financial companies – that wield overwhelming economic power in the modern world.

What has happened then – over the course of time and, if you will, quite *naturally* – is that property requirements for voting have been replaced by property requirements for running.  The result is about the same as it has always been – only those with property effectively determine who is elected to public office.  And if you – if property holders – control who can run and who can be elected, then, of course, you also control how, and in whose interests, they govern.

The Political Super Bowl and the Golden Rule:
Who's Got the Gold Makes the Rules

The electoral system, especially in the United States, basically parallels professional sports.  Two parties – for all practical purposes only two parties/leagues – are allowed to organize and run the sport – be it baseball, football, hockey or the Presidency – for the benefit of the owners and a few star players.  Just as for young athletes, an elaborate system of scouts and recruiters monitors the performance of young politicians at the local level, identifies those with promise – a combination of charisma, ambition and malleability – and provides them with the resources they need to progress through the minor leagues and on to the majors.  Those that make it to the majors are all similarly skilled and similarly trained, so that there is really very little difference between one team or one league and the other. But it makes for exciting and entertaining sport for the voting public to watch them play the game, and that keeps the public quiet and out of trouble while the riches just keep rolling in for the owners and a select few *superstars*.

In the words of Ralph Nader, "It's gotten so you can't tell Democrats from Republicans anymore in this country: They're both totally beholden to corporate America".  "Politics," he continues, "has been corrupted not just by money but by being trivialized out of addressing the great, enduring issues of who controls, who decides, who owns, who pays, who has a voice and access."  Those questions are better decided by the owners.

As was noted earlier, one-person one-vote may, in fact, be an excessively unrealistic standard.  Where very large disparities in wealth, education and access to information exist – in much of the Third World today as well as in rich countries like the US that have a high degree of inequality – the exercise of strictly equal suffrage may well contribute to the rise of demagogues and to much ideological instability.  So, it could in fact be wise to consider explicitly introducing some kind of a weighting system for voting, based on age and on educational attainment, perhaps.  This should be a topic of serious research, analysis and political discussion in coming years.

What could not be any clearer right now, however, is that the current situation in which the system is surreptitiously rigged by money is entirely unacceptable.  Unless it is changed, it will be impossible for government to play its necessary role as defender of the public interest, justly and democratically wielding its countervailing power to offset and control the massive, but undemocratic, strength of corporations and monolithic institutions like the professional military.  If government is not made able once again to play this countervailing role, then it is inevitable that the internal logic of unfettered materialism and the simple, universal, primordial drive of every organism and every organization to preserve and extend its power will soon lead us to catastrophe as the evils of extreme poverty, environmental destruction and societal disintegration grow beyond critical, irreversible thresholds.

The day is coming sooner, rather than later.

It is time to act.

## Beyond the Reach of the Law

In thinking about the proper role of government in using its countervailing powers to check the possible excesses of the corporate sector, it is important to keep clearly in mind that the modern corporation has, in many ways, already escaped the control of any one government. Modern corporations are truly *transnational* entities, operating in a myriad of jurisdictions and controlling resources that in many cases dwarf those of all but the largest national governments. Transnational corporations are able to effectively "shop" the world for legislation that is favorable to their narrow interests, especially in terms of generous tax treatment and the lax enforcement of environmental and labor laws. They are easily able to "break" governments that oppose their wishes, and, by pitting one against the other, force governments to compete for their favor in detriment to the interests of their citizens and those of other countries. Unless mechanisms are put into place soon to enable some kind of effective transnational cooperation between governments to alleviate the world's common critical afflictions – dire poverty, destruction of the environmental and our natural resources, and social disintegration and violence – any idea of governments individually exercising countervailing power and control over the destructive actions of transnational corporations is entirely ludicrous.

# IV:   What Can Be Done To Bring Change

So far we have been looking at what is wrong with our "system", and how it got to be that way.  That was the relatively easy part.  Although some of us have perhaps not yet been willing to look this closely into the workings of the system, perhaps not yet been willing to recognize the intrinsic character of its failings, nor perhaps yet been willing or able to see its various failings all at once and in relation to each other, there is nothing in the earlier sections of this essay that hasn't been said before.  What has been added is only, perhaps, the insight that normal self-interested behaviors sometimes can lead to highly undesirable outcomes, and that the "system" quite naturally produces – internally, as part and parcel of its workings – such evident and unacceptable evils as abject poverty, the destruction of our natural resources and the intensification of violence at every level of human existence.

Now that we know that it is our own actions that lead to these results, now that we know that things are getting worse, not better, and now that we realize that we have the means to do something about it, there is no longer any escaping our responsibility to act, consciously and deliberately, to bring an end to these evils before they bring an end to us.

We are responsible, and it is up to us to fix it.  We can do it.

Where and how to start?

Well, one approach, which is followed here, is to think first about those things that we can each begin to do immediately – that only depend on our own individual consciences, decisions and commitments – and that can begin to have an impact right away.  We may be discouraged by the thought that, singly, our individual actions may only be able to bring about small changes.  Maybe so.  Maybe not.  Certainly there are enough of us who are ready for change and who are willing to change our own individual behaviors, that, all together, we could have a major impact even in the very short run.

Think only of the possible impact of an election – coming up soon – that could bring truly fresh, clean and committed individuals into positions of power. That's an example of one big decision, made jointly by a lot of people. As we will see below, there are a myriad of decisions we face daily, that also have the potential to add up to something big.

So, let's give it a shot.

Secondly, we can begin to think about changes that may take a little longer to bring about, perhaps because they require new organizations to be built, and concerted legislative action to take place. These things may not produce major impacts right away, but could dramatically change our world within a period of five years or so. That's, of course, actually much sooner than it sounds, and to get that work done we also need to start making some basic decisions now.

Third are a number of things that would require more fundamental changes to take place, such as at the level of international cooperation and/or in making certain important amendments to fundamental commitments such as national constitutions, international treaties and charters. Here, we must of course be much more tentative, but it is possible to suggest some areas for discussion and analysis, and to encourage others to also begin thinking about fundamental changes that need to be brought about non-violently, and how to get started on that process.

It is presumed that these kinds of changes will generally take longer to bring about – maybe on the order of a decade or so – but this may not necessarily be the case. I think we will only find that out once we embark on the journey, look around, and see just how many others are also on the journey with us. My belief – my deepest faith – is that we are the multitude, and that together we hold the future in our hands.

What is very evident to me, however, is that any kind of meaningful and lasting change that we can bring about has to come from each of us individually first, and then spread non-violently through the exchange and voluntary acceptance of ideas, from the inside out, and from the bottom up. Historical attempts to bring about rapid change on a massive scale by forcing it from the top down or from the outside in have without exception ended in disaster.

60

Think only of the massive destruction and suffering undergone by humanity during the last century, as one visionary demagogue or another – of the right or of the left – sought to force rapid change through violent revolutionary means. Millions upon millions were slaughtered needlessly, social and economic progress stymied in vast regions of the world, for no good final result.

Violence can only ever be justified on the basis of self-defense against an imminent threat, and when no other means are available. Let us begin by accepting this lesson of history, and disciplining ourselves to greater patience and greater perseverance – and especially to greater faith in our own ability as human beings to develop and put into practice more rational and humane approaches to our own self-governance.

Step 1:  Grounding Our Beliefs

> Things that we can each begin to do immediately – that only depend on our own individual consciences, decisions and commitments – and that can begin to have an impact right away

What's the first thing we can do?

I've thought long and hard about the answer to this question and many possibilities have come to mind, but always I've come back to two words that ultimately mean the same thing:  "Love" and "God".

The first thing I think we all have to understand at the outset is *why* we seek to change ourselves and the world around us, even if it will take struggle – even if it hurts. It must be because we believe – deep down and each in our own way – that other people matter, that the world matters, and that there are things out there in the world that are more important than our individual lives and the satisfaction of our whims. There are things out there in the world that are awesome and unbelievable, that leave us awe-struck and dumb when we think about them. Things that are "sacred".

One of them is Love.

Love of parent, love of child.  Love of spouse.  Love of country.
Love of God.

Where does that all come from, and why?

The desperately-scientific among us may respond, "Love is an
adaptive emotion (whatever that might be) that builds cohesion
between members of a reproductive group of individuals, and
thereby adds to their ability to survive".

Oh, really?  It may well be all that, but it is obviously also so
much more, as we can almost all attest from personal
experience.  And it is certainly not a necessary emotion for
survival, as the success of any number of species also attests.
The bacteria and the cockroaches, alligators and roseate
spoonbills are all doing just fine, thank you, without any
observed tendency to drop their lower jaws in awe and wonder.

So what is love, and where does it come from?

Who knows?

But, it's there.  A part of us, just as we are part of the world
around us.

Where does the idea of beauty come from?  Why, again, that
sense of awe and wonder when witnessing a light summer rain,
a spectacular sunset or an image of a distant cloud of galaxies?
What good does that do?  What is it for?

Who knows?

But, it's there.  Undeniable and breathtaking.  A source of
meaning.

And what of life itself?  Where does that come from, and what is
it for?

Who knows?

But, marvel of all marvels, it's there.

You're there.  I'm here.

What is that all about?  What for?

Who knows?

But, we can speculate.

One very famous and quite intriguing speculation is due to Nobel-prize winning Austrian physicist, Erwin Shrödinger, one of the pioneers of quantum theory and a contemporary and friend of Albert Einstein.  He published a book in the mid-1940s entitled What Is Life? And Other Theories in which he defined life as *negative* entropy.  Entropy, as stated in the Second Law of Thermodynamics, is the tendency of all closed systems in the universe to move unidirectionally from higher to lower energy states, and finally to some kind of ultimate quiescence.  Living things, observed Shrödinger, exhibit the opposite characteristic at least locally, moving from lower to higher states of organization and energy, in part perhaps by absorbing energy from the surrounding environment.

While clearly much too narrow a concept to come close to *defining* a phenomenon as big and as diverse as "life", negative entropy is at least one instance in which physicists acknowledge, but of course can not fully explain, the existence of some kind of organizing force within the observable universe.  In fact, right smack in the center of the *observing* (perhaps self-observing) universe, manifested by the thoughtful scientist himself.

Yes, life is part of the universe.  It is the part we know most intimately, but perhaps know the least about.  It is everywhere that we know on Earth, and probably in the Universe beyond. We living things are part of the universe, and we humans are part of that part which also appears to have the attributes of consciousness and self-consciousness.  But life extends far beyond ourselves and is pretty darn hard to define when you come down to it, going far beyond what we have allowed the physical scientists to define for us.

One such standard definition, from one of my favorite sources of information, the Wikipedia, is as follows:

Properties common to these (living) organisms—plants, animals, fungi, protists, archaea and bacteria—are a carbon- and water-based cellular form with complex organization and heritable genetic information. They: (1) undergo metabolism; (2) possess a capacity to grow; (3) respond to stimuli; (4) reproduce; and, (5) through natural selection, adapt to their environment in successive generations.

An entity with the above properties is considered to be a *living* organism, that is an organism that is alive hence can be called a life form. However, not every definition of life considers all of these properties to be essential. For example, the capacity for descent with modification is often taken as the only essential property of life. This definition notably includes viruses, which do not qualify under narrower definitions as they are acellular and do not metabolise.

Neither do viruses reproduce by themselves, but can only reproduce by entering the nuclei of cells and replacing the cell's DNA with their own, forcing it to reproduce the virus instead of itself. The extent to which viruses "grow" (as opposed to replicate), and the extent to which they respond to stimuli, are also questionable. So that, in the case of viruses, we are already close to a vanishing point in the differentiation between "life" and "non-life" in the microscopic direction. This is even more dramatically the case with prions – the very weird small things that can infect us and other species and are believed to cause "mad cow" and Creutzfeldt-Jakob disease, and which do not even possess nucleic acid and do not "evolve".

Infectious particles possessing nucleic acid are dependent upon it to direct their continued replication. Prions however, are infectious by their effect on normal versions of the protein. Therefore, sterilizing prions involves the denaturation of the protein to a state where the molecule is no longer able to induce the abnormal folding of normal proteins. However, prions are generally quite resistant to denaturation by proteases, heat, radiation, and formalin treatments, although their infectivity can be reduced by such treatments.

What the above is saying (read "kill" instead of "denaturation") is that prions are pretty hard to kill. Indeed, the Wikipedia article cited goes on to say that you can "denature" prions by subjecting them to high temperatures under pressure for extended periods of time, but then also notes that under certain circumstances it has been possible to "renature" denatured prions, or, in some sense, to bring them back to "life".

Proteins, scientists have also speculated, can be formed "naturally" (how else, I wonder, could they have gotten here) by sparking electricity over a molecular soup in a methane-rich environment (or something like that) much as is supposed to have existed during the early history of Earth.

The point for me is that – in the micro direction – it seems to be clear that there is no "definite" boundary between life and non-life, and that – it would appear – all material beings, including ourselves, relate to one another along a single continuum and may, in fact, all be part of "the same thing".

There are also interesting speculations that one can indulge in in the macro direction. One of the most fascinating things astronomers are able to look at and talk to us about is the "life" of nebulae, stars, and galaxies in the observable universe. Here again, instances of movement in the direction of higher states of organization (of negative entropy) are to be found everywhere, with all manner of astronomic objects continually coming into being and evolving in some kind of eternal cosmic balance with the entropic force. Equal and opposite and seemingly eternal.

Taken as a whole (whatever that means in the context of infinity) it seems we would have to say that the universe constitutes a "closed system", that there isn't any "other" place from which it could be drawing energy. Yet it clearly manifests negative entropy, everywhere and, apparently, always has.

Is the universe "alive"?!

I wonder.

Seems that way, in some sense.

And it is all one, isn't it? Including us.

Why is all of this important, and being brought to mind in a book that is about the defects in our economic and political system, and things that we might be able to do to correct them?

I believe that all of this is important because, as individuals looking for the vision and courage needed to rouse ourselves to action, and for the strength to persevere in that action, we need to ground our convictions in a moral framework that "is larger than ourselves" and capable of being conceived of as "universal" and "true". We also need to know how far we can go, what are the limits that we must impose on our own actions, and where they come from.

It used to be that our forefathers believed that that vision and courage, that framework and those limits, came from their God and their religion.

That's Ok. That was a pretty good way to look at it at the time.

But, unfortunately, and maybe this is just another way in which the modern system[5] seeks to isolate us from one another and rob us of our power to act independently, for many of us, including me, conventional notions of God and the tenets of traditional organized religions are no longer able to compel our imaginations, our credulity or our faith. Unable to believe in the religions of our fathers, we must either explore our lives and the world ourselves looking for alternative, more credible faiths, or be satisfied with an empty existentialism that justifies nothing and everything at once.

The first step, I think, in being able to make changes in our own lives that can help to fix the system, is to find our own moral and metaphysical basis for action and belief. Each of us needs to undertake our own search, though we can of course compare notes and share our experiences and insights along the way.

---

[5] Unlike that of our forefathers, who drew strength and legitimacy from their conceptions of divinity and "natural" law which said, for them, that each of us individually has worth, and that we have been "endowed by our Creator with certain inalienable rights, to life, liberty and the pursuit of happiness..."

That is the first step we must all take, and that many of us have been discouraged from taking, in order to begin a process that can have the power and the strength to bring about the kinds of changes that must be made if we are to change the course we are on, and help to bring about a fair, sustainable and joyful existence for all of us sharing in the adventure of life on Earth.

I, for one, have come to believe that all of life and all of existence, for all time and in all places, is "God".

"Everything" = "God"

"Everything" is marvelous and "good", a gift, that simply "is".

Like regions of high entropy in the physical universe, evil is either an incomplete perception of reality (not really evil, just necessary, as in death), or simply the localized and temporary "absence" of "good" (of God, of love, of life, of beauty...).

We are part of "God".

Though "God" is not, I don't think, anything like a "person", "God", at least to the extent that we exist and are "conscious", appears to have some form of consciousness.

We are, it appears, one of the conscious parts of God.  As part of God = Universe, we are part of the self-conscious part of the (living) Universe.

By our actions and our choices, we are able to make evident – to manifest – the existence of God in the Universe and, indeed, the very equivalence of God and the Universe.  Ultimately, we can flow with and affirm the life force – and enjoy a higher, more blissful level of consciousness – or ignore our own lives and our own reality, our own miraculous participation in the marvel of "Creation".

It's more a question of attitude and the conscious exercise of will than anything else.

But the above, and nothing more – just the celebration of the undeniable and marvelous existence of "life", of "love" and of "creation" – is enough, in my case, to give my life a sense of purpose (most of the time), to guide my actions in an affirmative sense, and to show me where the limits are.

In particular, I am led to seek to serve "God" in my life, and to help make manifest the wonder and the glory of life and existence. I am led to fight my own greed and selfishness to the extent that I am able. I am invited to seek love, and to expound hope. I am obliged to renounce violence in all of its forms as a means of fulfilling my own ambitions or desires.

I have been humbled by the wondrous gifts I have received, and I am thankful for it. I seek opportunities to make restitution however I am able.

That is my faith. That is what leads me to try to make things better.

Affirm Love. Find God, and learn to love God[6]. That is the first step that I believe that you and I must take before we come together.

Step 2:  Coming Together

After securing the moral framework that you need, the next step must clearly be the seeking out of other people who agree only in that something effective must be done to eliminate the three great evils that our current system not only does not seem to be able to address but, indeed, seems to actively perpetuate.

After coming fully to grips with our shared reality and understanding where it come from and how it works, the next step must clearly be the seeking out of other people who agree that something effective must be done to eliminate the three great evils that our current system not only does not seem to be able to address but, indeed, seems to actively perpetuate.

---

[6] Note that I have not said, "Find religion". Religions in the course of human history have by and large resulted in the politicization and exploitation of people's faith and beliefs, with many tragic consequences. Secular "religions" are just as dangerous and have historically led to equal or greater excesses, as in the humanist revolutionary movements, all motivated by the quest for social justice, that led to so much slaughter during the 20th century. So, let your faith guide your politics, but keep faith and politics apart.

No one can accomplish anything meaningful alone.  We must come together to exchange our views, consolidate and strengthen our ideas and pool our resources in support of a program of action.

There are many ways to do this, especially with the astounding personal communications technologies that are now available to us.  Many sites on the Internet offer people the opportunity to form interest groups of all kinds, and to share information on a scale and scope that has never been seen before.  Used creatively these will be of assistance in providing multiple avenues for communicating with one another and gradually building the organizations – planning committees, action/project committees, think tanks, publications, new alternative political parties – that will be needed to give substance and reality to the changes that we want to bring about.

This process of coming together must begin locally.  It is not enough to just establish virtual contact.  It is also necessary to be able to come together physically in groups from time to time, to be able to have more open and free-ranging exchanges, and to be able to build the personal bonds and the commitments that will give force to the movements of the future.

Step 3:  Putting Our Beliefs into Action

Beyond the all-important step of reaching out to others and beginning to organize into discussion and action groups, there are many, many things that we as individuals can do to begin to exert pressure on the system.

Many of these things are already being done, by many people.  We can join them, and make this movement stronger, more quickly.

To begin with, we can change the way we spend our money.

So much money, in the United States and Europe and Japan and among the affluent all around this crazy world is wasted on nothing more than hedonism and self-gratification, on ostentation and the relief of boredom.  This is pathetic and can be changed by you and by me, by all of us, since to a greater or lesser extent, we are almost all guilty of some degree of wasteful self-indulgence.

Forget about Las Vegas and Cancún.  There is nothing happening there that matters anyway, so let it just stay there, and rot.  Go to Sri Lanka or Nepal.  Go to Sierra Leone, Ghana or Tanzania.  Go to Nicaragua or Peru.  For that matter, visit a community group in a neglected neighborhood in your own city, or visit an Indian reservation in the American Southwest.  Get to know some real people and see how they live.  Catch some of their joy, admire their fortitude, realize that they are your brothers and sisters, and understand that their pain must also be your pain.

Buy a Chevy or a Toyota instead of a Hummer or a Mercedes, and put the difference into social investment.  There are these days any number of vehicles for putting some of your excess money to work in socially productive ways.  You can even go online (see www.kiva.org) and directly select among thousands of applications from community groups and small enterprises around the world, and make them a loan for some productive purpose.  Chances are you will be repaid in full, with interest and with love.  Any number of alternative social investment groups operating around the world, and their contact information can be easily found on the Internet.

Buy products bearing the FairShare Foundation logo.  This logo signifies:  (1) that the company making the product that carries the logo has self-certified that it adheres to a high standard of social and environmental responsibility; and, (2) that 5 cents out of every dollar the manufacturing company or brand-owner receives at wholesale, is contributed to a social investment fund managed by the FairShare Foundation.

The FairShare Foundation (see www.fairsharefund.org), which has been incorporated in the State of Maryland and which has been recognized for tax-exempt status by the U.S. Internal Revenue Service, was set up by me and a small group of economic development professionals to manage a social investment fund which can finance small income-producing projects in disadvantaged communities anywhere in the world, with the aim of helping recipients build a small business or otherwise improve the economic condition of their community.

Ultimately, funds provided by FSF are expected to be repaid and so to replenish what is hoped will be an ever-expanding pool of resources available for productive social investment purposes around the world.

If it hasn't already been done, somebody please establish a *Consumer Reports*-type of online publication and reference site focused on evaluating and rating the environmental and social performance of major corporations and their products. *Multinational Monitor* does some of this work, but has a relatively limited "readership". Maybe *Consumer Reports* itself would be interested in broadening the scope of their evaluation and rating activities and very likely would be in the face of strong evidence of potential demand for such a service. Send them an e-mail, and let them know what you think.

In the meantime, websites such as www.care2.com and others like it listed on the Internet provide a lot of useful information that will help you make better consumption choices. Even if you can't be 100% consistent, you can reduce the amount of your money that goes to countries with poor labor and environmental norms – such as China in particular – and to companies with a record of disregard for these issues.

Almost everybody owns some stock these days, either directly or through a mutual fund or retirement plan. If you don't already know, find out what companies are in your investment portfolio, and learn more about what they represent. Most corporate websites these days include an "investor relations" page, where annual reports and other pertinent information can be viewed and/or downloaded. In the U.S., the Securities Exchange Commission (SEC) publishes all mandatory filings submitted by listed corporations, and these can be found at the following address:

www.sec.gov/edgar/searchedgar/webusers.htm

If your share ownership is through a mutual fund or pension plan, you can contact the fund manager for information on the exercise of your voting rights. Also, if there are issues involving the activities and policies of a particular corporation that merit the effort in your mind, you and a group of like-minded individuals can buy shares in that company, and become active in the affairs of that company through participation in shareholder meetings, making petitions to management, and circulating communications to other shareholders.  Of course, such activity undertaken by you and your group within the structure of the company's internal communications and voting procedures can also be publicized by recourse to the media and the web in an effort to build additional support for your positions.

Although a detailed analysis of necessary reforms of corporate governance laws is beyond the scope of this book, we can certainly point to immediate actions that are open to individuals now to work, as it were, "within the system" to bring about some change in the right direction.  The reform of corporate governance rules and practices will also be highlighted below as a priority area for research, analysis and legislative action in the medium-term.

In the spirit of beginning to do something *now* even it that can only be "within the system", you should join and participate in an existing or new political organization in your locality, as this will give you the opportunity to exchange views with other concerned individuals and build a platform that will allow you to effectively pressure the existing political structure from the bottom-up for the changes that most directly affect you.

An Agenda for Collaborative Action over the Medium-Term

While there is a great deal that can be done at the individual and small group levels, deeper legislative and institutional change will require that individuals and small groups develop and refine their individual platforms, and find each other or find other pre-existing larger organizations that will permit them to massively advocate for the changes that are contained within their platforms.

We can't do more than suggest a few ideas and directions at this point, which we attempt below.  Though it should only take a few months to get the process started, it will take time for ideas to "percolate", mature and develop, and for groups to come together to form their organizations and a common, broader agenda that can be supported by a large number of people. Keeping the objective of ridding ourselves of the three great evils that currently characterize our system clearly in mind, we must:

- stop the regeneration and perpetuation of poverty
- stop the destruction of our environment
- stop violence and the disintegration of the human community

Key items suggested for inclusion in a broad-ranging action agenda for the medium-term include the following.

### Really Leave No Child Behind

Supposedly side-tracked by 9/11 and everything that has come since, the Bush Administration's "Leave No Child Behind" education program was one of the most cynical political swindles ever perpetrated on the American people. This program, put forward as the heart and soul of a new, "compassionate conservatism", has turned out to be nothing more than flim-flam play-acted by that folksy pair from Crawford, Texas whose continual deception of the American public, bafflingly successful in light of their evident shallow and venal ineptitude, only just began with "LNCB".

Thankfully, they will be gone soon.

For now, what is important is to remedy the damage that was done, and to make good the promises, putting forward real programs that can make sure that the Bush Administration is remembered as the last to ever leave a generation of American children behind.

What must be done?

Develop and fund private and public programs at the local, state and federal levels that will *ensure* – with no further excuses tolerated – that *all* children in America have access to adequate nutrition, good health care and a good functional and moral education through the high school level. The process of developing and implementing such a program must be both bottom-up – involving fully all the actors who must play a part at the grass-roots level – *and* top-down, with vision, leadership, conflict resolution *and* funding being provided by a new President and Congress.

Making "Leave No Child Behind" a reality in America will probably cost less than one month's spending on weapons and warfare at their current levels[7]. Not only can we make sure that all children (yes, also including the children of illegal "aliens"[8]) in the United States get an equal and fair chance to grow into healthy, happy, productive and responsible adults, but – I can assure you – the investment made today will be recovered many times over in a happier, healthier and more productive society tomorrow. I'm talking money here, not "just" human lives.

It's only a question of doing it, instead of lying about it. It can begin in 2009 and become a *sho'nuf* reality by not later than 2010.

Yes, we can. For the most part, Europe and Japan already have.

### *Expand Funding for VISTA and the Peace Corps*

Bringing adequate nutrition, a basic standard of health care and at least primary education to all children in other parts of the world will take longer and will require overcoming more difficult obstacles. But one thing the United States and other rich countries can do relatively quickly is to expand political and financial support for programs such as Vista and the Peace Corps. Young people from the First World, volunteering their time and skills to help others that are in need in their own countries and around the world can accomplish amazing results at relatively low cost. At the same time, these big-hearted young people can gain real-world experiences that will open their eyes and expand their minds as well, and shape their

---

[7] Suppose, as a pretty extreme scenario, a total of 5 million very poor children in the U.S. each requiring assistance through public investment of $10,000 per year – we are only talking about something like $50 billion, a pittance compared to current U.S. military spending of close to $700 billion annually. If necessary, we should be willing to spend several times this amount, and the source of the necessary revenues should be obvious.

[8] These kids may someday have to be deported as a consequence of their parents' transgressions (let's hope not), but until that happens, they, like every other child in America, deserve an even break. If nothing else, to save ourselves problems down the road, but really for much more than that.

attitudes for the remainder of their lives. There can hardly be any better preparation for good citizenship and a full, productive life than spending a couple of years sharing one's gifts with other people, and such service should be rewarded by the rest of society with the same preferences in education and other areas that are currently given to young people in military service.

*Remove the Most Glaring Distortions from Our System*

Other relatively easy steps that can be taken quickly to reduce the cost of basic services and improve the standard-of-living of our people would include, for example:

- reform medical malpractice laws to establish reasonable ceilings on jury awards to victims, not out of insensitivity to victims, but as a practical measure to rapidly reduce medical malpractice insurance costs, and, consequently, the cost of health insurance and medical services to the public. It would be important, among other issues to be addressed in the context of amending current legislation, to focus specifically on the way the current insurance/malpractice/profit-based medical system perversely but inexorably leads physicians to order much more expensive laboratory and diagnostic testing than is often needed
- extend the period of unemployment insurance coverage for as long as people searching for new employment are willing to work part-time at community-based public service jobs
- analyze the causes behind the exorbitant increase – comparable to that of healthcare – in the cost of higher education that we have observed in recent years. Study the need to perhaps require all institutions of higher learning that receive any kind of public support to provide a higher minimum percentage of full-time scholarships to lower-income students. Reward public service by young people with scholarships/financial assistance so that they can continue their studies through the post-graduate level, and reduce the interest charged young people on their student loans. Any public resources invested in reducing or eliminating interest on student loans would pale in significance in comparison with our current levels

of military spending, and will provide massive economic returns to society at large

- end the harassment of small businesses by the Internal Revenue Service, which currently imposes high costs on those businesses that make an effort to comply with excessively frequent and complex reporting requirements, and punitively high penalties and interest on companies who may not owe any tax but fail to deposit or report taxes on time due on the very strict schedules that the IRS imposes. Have the IRS focus seriously on closing loopholes and illegal tax evasion by large corporations and high-income individuals.
- eliminate or sharply reduce "corporate welfare" payments/exemptions allowed by different levels of government to large corporate interests in the agribusiness, mineral, fishery and forest-products industries. In the U.S., these amount to *half-a-trillion dollars* or more, annually.
- revise legislation pertaining to corporate governance and minority shareholder rights, including voting rights. Key issues include: corporate social responsibility, labor practices, environmental safeguards, executive compensation policies, and international outsourcing.

Effective Global Environmental Management

It is one Earth, and we share one, seamless natural environment. All of us alive today, and all of our descendants for generations to come depend on a common pool of natural resources – air, water, soil, plant and animal gene pools, mineral deposits, etc. – that, degraded or destroyed in any one place, degrade or destroy our "common wealth" – the natural wealth of this Earth that we have all inherited equally – perhaps forever.

The limited legal and regulatory frameworks that we have so far been able to put into place to protect our environment, natural resources and "common wealth" have largely been instituted and are almost totally enforced within the frameworks of individual national legislation. Individual, national legislation and enforcement is nowhere near being able to accomplish the task. That is one of the main reasons that we have been so ineffective up to now in reversing an accelerating trend of environmental and natural resource degradation at a global level.

Profit-seeking multinational corporations, as they will be the first to tell you, are not in business to save the world. They are in business to make as much money as they can, in most cases more or less within the limits of "the law". The problem is that individual national laws are very pliable, and very unevenly enforced. So, in their presumably justified quest for maximum profits, multinational corporations – many of which are far richer and more powerful than all but a handful of national governments – shop the world for the most lenient possible regulation of their activities insofar as their impact on the environment and our common pool of natural resources is concerned. Poor countries around the world – even relatively poorer states and provinces within more prosperous countries – are so desperate for investment and jobs that they are many times willing to dilute their environmental legislation almost to the vanishing point, and to allow what few laws and regulations they may need to keep on the books for public relations purposes to go totally without enforcement.

If one jurisdiction tries to raise their standards or put teeth into their enforcement efforts, other jurisdictions will take advantage in an effort to lure the investment and jobs provided by the large multinational corporations to their localities. So it becomes – very evidently – a headlong race to the bottom, and our lives and the future of succeeding generations are being threatened as a consequence.

Yet, there is no innate evil intent involved, necessarily. It is just the evil result of the internal dynamics of a system whose structure is no longer up to the task of sustaining life for all of us who share the planet.

The only way to countervail the power of multinational corporations is to set up international governmental structures that can negotiate minimally acceptable environmental standards on a global level, and establish the mechanisms and generate the resources necessary to ensure a uniform, coherent standard level of enforcement everywhere around the world. If this is done, nowhere will any corporation competing internationally be put at any special disadvantage, and the incentive for these corporations to shop the world for the weakest regulation and enforcement will come to an end. The costs of implementing a common higher standard of environmental protection around the world will, of course, be passed along to consumers, and that is Ok, as we, ultimately, are the parties who have an interest in making sure that our common wealth is preserved and passed along intact to our children and theirs.

The United States has so far missed a major opportunity to take leadership in the development of a common global standard of environmental protection through its failure to broaden the international dialogue and action agenda beyond the issue of global warming. A more responsive attitude on this subject, which does not mean the abandonment of US demands for fair treatment, universal adherence and reciprocity, could allow a range of other pressing issues – involving toxic and human waste, agricultural runoff, and the depletion of world fisheries, for example – to also be included as priorities for urgent action at the international level. As concerned citizens from whatever country, we should take each and every opportunity that is open to us to act and advocate for strong, effective international cooperation to establish and enforce a universal standard of environmental protection.

Protecting Workers' Rights around the World

A corollary issue is, of course, the treatment and rights of workers around the world. In the same manner as the deficiencies of the current system almost force multinational corporations to shop – even to encourage – lax enforcement of environmental protection, so do the deficiencies of the system, without presuming any evil intent on the part of multinational corporations, practically force them to shop the world for the lowest wage and the most lax protection of labor laws. Can it be any surprise, then, that the abuse of workers – mostly women and children – even to the point of virtual enslavement continues to be so prevalent in so many countries around the world?

The disparate state of economic development in these countries, along with their different and unique histories and cultures, makes the issue of more equal treatment of workers extremely difficult and complex.  Some countries are so poor and so desperate for any employment that at least allows for subsistence from one day to the next, that it would be difficult to press for overly rapid change without threatening the very survival of large numbers of workers.  Some paid work, no matter how meager the payment or difficult the conditions, may be better than no work at all.

Labor conditions and treatment in different places around the world are, of course, also directly related to international labor migration – both legal and illegal – which poses major challenges within rich nations to develop humane *and* practical immigration policies.

It would be naive to expect very rapid progress in developing a coherent international approach capable of redressing the evident inequities in the treatment and remuneration of workers in different parts of the world.  However, it is crucially important that we recognize that the system is failing badly in this regard, and that we must – through advocacy, exchange, negotiation and the gradual construction of consensus – begin to address the issue as a critical priority at both the national and transnational levels.

There is no way that any single country can solve this problem on its own, but, so long as the problem is not solved, there is no way either to expect international corporations to do anything more than to continue to respond to the adverse and deeply perverse incentives that are currently built into the system.

## Get Out of Iraq:  Build a Non-Military Foundation for International Peace and Security

We've got to get out of Iraq, as soon as possible.

What Bush says is true.  The presence of U.S. troops there is a magnet that attracts insurgent elements from throughout the region to fight against us, and has resulted in the expansion and extension of what looked like a police action to secure the country just after the American invasion, to a major conflagration that has resulted in the needless loss of thousands of American soldiers, and tens of thousands of Iraqis.

The war itself was unnecessary and uncalled for. Whatever the real motives of the Bush Administration, *there was no credible evidence of the existence or of preparations to manufacture Weapons of Mass Destruction in Iraq!* And the doctrine of preemptive military attack to prevent terrorism, supported solely on suspicion of possible conspiracy against us, is untenable on any national or international legal grounds or precedent in the history of civilized countries. *War kills people, including innocent children and their parents, no matter how "smart" the bombs! War should always be held back as long as possible, as the absolutely last resort, when every other option has been tried and failed.*

*Good intelligence*, as will be argued more extensively below, has been and continues to be our best defense against global terrorism. *Justice, good policy and good statesmanship* are our best offense in winning the hearts and minds of people around the world, people without whose support terrorism has no chance of success against us.

It is easy to paint terrorism as the "mindless" acts of religious or political "fanatics", who are impossible to understand or communicate with. It has always been important to dehumanize an enemy as much as possible. It makes it much easier to kill them that way, and for otherwise civilized citizens to witness and accept the carnage as unavoidable. That way, when our "smart bombs" falling out of the night all of a sudden transform a once happy eight year-old into a paraplegic orphan, without parents, without siblings, without arms and without legs – without any illusions about what life holds – we can feel sad instead of feeling guilty. That is much easier.

### It Is Our Own Whirlwind That We're Reaping

But it is not so difficult to understand that terrorists are nothing more than other, slightly different, but basically misguided humans, just like us. Al Qaeda were once our allies, fighting, with our help, a holy war against the godless communists in Afghanistan. Together, we beat the godless communists, and forced them to retreat back into Soviet Russia. Then, of course and despite whatever promises we may have made regarding postwar reconstruction assistance, we walked away and left them to their own devices – and the Taliban.

Another ally of that epoch – perhaps emboldened by the strength of his former ties to the U.S. during the bloody fight against the diabolic ayatollahs – Saddam Hussein launched an invasion to reclaim an oil-rich former province of Iraq.

A despot much like all the other rulers who came into power in the Middle East in the decades following the fall of the Turkish Ottoman Empire – sustained and supported by the British and American oil companies and clandestine agencies – Saddam had achieved some measure of distinction and appeal within the region as one of the few secular, non-monarchic rulers in the region. As such, he was a definite rival and a threat to the ruling house of Saud in Saudi Arabia. As such, he was a definite threat to Western oil supplies lying beneath the sands of Saudi Arabia, and "guaranteed" to us by the house of Saud. So, we established massive bases in Saudi Arabia, and, with the blessings of the "desert kings", we launched a massive attack against Saddam's forces in Southern Iraq and Kuwait.

We won, with huge loss of life on the Iraqi side, but relatively little on our own.

That was good.

The more experienced elder Bush realized that conquering and attempting to rule over Iraq would lead us into a quagmire to no good strategic purpose. So he ended the hostilities following the liberation of Kuwait, and withdrew American and coalition ground forces from Iraq.

That was also good.

What was not so good, from the perspective of our former allies the Al Qaeda "freedom fighters", was the elder Bush's decision to establish permanent U.S. military bases in Saudi Arabia and Kuwait. That, despite the apparent acquiescence of the royal house of Saud, smacked too much of "occupation" for those who had just spent a decade in the expulsion of other Western infidels from a neighboring Muslim country.

While we can neither accept nor condone terrorist violence against us or any other human beings anywhere, we can attempt to understand the motivations of those who fight against us. In the case of the Muslim revolt against the West – spearheaded at the moment by Al Qaeda – the factors against us are fairly clear.

One, the imposition and maintenance in power by the West of corrupt, backward and despotic regimes throughout the Arab world – which includes most of the former Ottoman Empire that extended from the Balkans in Europe to Morocco in Northern Africa, fundamentally to allow the West to exercise control over Middle Eastern oil supplies

Two, the establishment of Israel[9] in 1948 within the British Mandate of Palestine, and the expulsion of over 700 thousand

---

[9] A violent process in which Jewish "terrorists" including future Israeli Prime Minister Menachem Begin were heavily involved. The following passage from the Wikipedia encyclopedia on the life of Begin is illustrative: "Begin issued a call to arms and from 1944–48 the Irgun launched an all-out armed rebellion, perpetrating hundreds of attacks against British installations and posts. Begin financed these operations by extorting money from Zionist businessmen, and running bogus robbery scams in the local diamond industry, which enabled the victims to get back their losses from insurance companies. For several months in 1945–46, the Irgun's activities were coordinated within the framework of the Hebrew Resistance Movement under the direction of the Haganah, however this fragile partnership collapsed following the Irgun's bombing of the British administrative headquarters at the King David Hotel in Jerusalem, killing 91 people, including British officers and troops as well as Arab and Jewish civilians. The Irgun under Begin's leadership continued to carry out military operations such as the break in to Acre Prison, and the hanging of two British sergeants, Clifford Martin and Marvyn Paice, causing the British to suspend any further executions of Irgun prisoners. Growing numbers of British forces were deployed to quell the Jewish uprising, yet Begin managed to elude captivity, at times disguised as a rabbi. The British Security Service MI5 placed a 'dead-or-alive' bounty of £10,000 on his head after Irgun threatened 'a campaign of terror against British officials', saying they would kill Sir John Shaw, Britain's Chief Secretary in Palestine. An MI5 agent codenamed Snuffbox also warned that Irgun had sleeper cells in London trying to kill members of British Prime Minister Clement Attlee's Cabinet."

As the Israeli War of Independence broke, Irgun fighters joined forces with the Haganah and Lehi militia in fighting the Arab forces. Notable operations in which they took part were the battles of Jaffa, Haifa, and the Jordanian siege on the Jewish Quarter in the Old City of Jerusalem. One such operation in the Palestinian village of Deir Yassin in April 1948, which resulted in the death of more than a hundred Palestinian civilians, remains a source of controversy.

Palestinians from Israel during the first Arab-Israeli War in 1948-49, who, after 60 years, have yet to recover a fully independent homeland.

Three, the cultural invasion of the Muslim world by what are viewed as shallow and corrupt Western values that threaten the integrity of the Muslim family and the stability of the Muslim community. This includes, at the top of the list, what is viewed as the perversion of the traditional primary role of women in society as wives and mothers, transforming them for commercial reasons into pleasure-seeking libertines, with themselves and society the victims of unfettered capitalist exploitation of promiscuity and pornography.

The Muslim revolt against the West is primarily driven by a sense of victimization by the West, and by one of relative impotence that only leaves open the alternative of desperate measures and methods.

Unlike the drug lords who perpetrate a vastly more massive and profound terrorism against the children of Western society, the Muslim suicide bombers are mostly devout and clean-living. In their eyes, their actions are defensive, a sacrifice made in defense of their homelands, their culture and their beliefs. Much like our own soldiers who have been willing to sacrifice their own lives to save others by throwing their bodies over a hand-grenade, or by storming a machine-gun nest to kill the enemy, for example, these young men and women are viewed as heroes by most of their contemporaries in the Muslim world today. And if, in the course of their necessarily asymmetrical attacks against those whom they believe to be their enemies they also inevitably kill some innocents, their acts are, in their view, no more heinous and no less inevitable than the acts of those who launch missiles and aerial attacks against their communities, killing innocents – "collateral casualties" – who, in the long and

---

Some have accused the Jewish forces of committing war crimes, while others hold those were legitimate acts of warfare, however it is generally accepted that the Irgun and Lehi forces who took part in the attack carried out a brutal assault upon what was predominantly a civilian population. As the Irgun's leader, Begin has been accused of being responsible for the atrocities that had allegedly taken place, even though he did not partake in them.

cruel course of the various Arab-Israeli conflicts and the two U.S.-Iraqi wars, by now must number in the tens of thousands.

It is all highly regrettable, on one side as on the other. Though there have undoubtedly been human monsters involved on both sides, they, as a people, are no more monstrous than we are. We can understand one another if we want to and if we try to.

It is way past time to stop the killing. Let's stop it now.

While any withdrawal needs to be executed with professionalism and care to avoid any further needless losses to our forces and maximize the probability of peace ensuing, by taking away a primary pretext for violent resistance an American withdrawal from Iraq will do more than anything else I can think of to bring stability to the region. It will permit the government of Iraq to request the return of expatriate combatants to their homes, and close down what has become the world's greatest training ground for insurgency and terror. There can be no doubt that the United States will find ample support and cooperation from other Western nations and international organizations to ensure that the transition to peace in Iraq takes place as uneventfully as possible.

### Living by the Powell Doctrine

Preemptive military action is not the way to world peace, or even to bringing about even a small measure of short-term security for the people of the United States and other Western nations.

A much more reliable guide that should limit the use of military means to only the most necessary and unavoidable cases is attributed to former U.S. National Security Advisor, Chairman of the Joint Chiefs of Staff, and Secretary of State, General Colin L. Powell. The questions posed by the Powell Doctrine, which should *all* be answered affirmatively before military action is initiated, are:

1. Is a vital national security interest threatened?
2. Do we have a clear attainable objective?
3. Have the risks and costs been fully and frankly analyzed?

4. Have all other non-violent policy means been fully exhausted?
5. Is there a plausible exit strategy to avoid endless entanglement?
6. Have the consequences of our action been fully considered?
7. Is the action supported by the American people?
8. Do we have genuine broad international support?

In the case of Iraq, it is questionable if even one of the Powell criteria were met.  The American populace was deeply traumatized by the events of September 11, 2001 and may perhaps for this reason be forgiven this lapse.  But it should never happen again.

General Powell served the United States with high honor and distinction for several decades.  It is unfortunate that, as Secretary of State in the first Administration of the second Bush, he felt compelled – whether through loyalty and obedience or personal ambition – to himself betray the principles for which he is most likely to be finally remembered in world history.

What alternatives were there in the case of Iraq, and what alternatives are there to deal more constructively with future challenges?

One such alternative is the organization and execution of international policing actions when sanctioned by a duly constituted international body and declared to be necessary to head off a perceived threat from rogues such as Saddam Hussein.

In his case specifically, enough evidence existed of crimes on a massive scale against his own people to have justified his indictment by the International Criminal Court (to which the U.S. unfortunately does not belong) or by some special *ad hoc* committee that could have been created to review the evidence by the U.N. Security Council.  Such an indictment could have ordered the surrender or capture and arrest of Hussein and a small number of complicit henchmen.  Backed by the threat of force to accomplish an arrest and perhaps further encouraged by a monetary reward, it is more than likely that Saddam would have been betrayed by his own cronies within a short period of time, captured (or killed) and turned over for trial by international authorities. Had that happened, Saddam would today be languishing in a jail cell somewhere, much like Manuel Noriega, and tens of thousands of people would still be living.

Even in the worst case scenario, in which his bodyguards and military forces remained loyal, it would surely have been possible to organize a large-scale policing action to pinpoint his whereabouts and accomplish his arrest or elimination – with the sanction of a duly constituted international body. With control of Iraqi skies brought about with little loss of life on either side, a United Nations policing force of no more than 5,000 men could have rapidly accomplished the extraction, withdrawing immediately from Iraq thereafter, and leaving the country and its people to find their own way towards peaceful coexistence and development, free of the maniacal and criminal despot who had taken power over them.

This is not an impossible scenario. It would very likely have worked. The world would be a very different – and safer – place today, and we would have put into practice a new model of international cooperation to deal with future major threats to international security and peace.

Beyond the development of alternative means for deploying force without creating war, the U.S. and other leaders among the international community need to develop a whole array of preventive instruments and measures that can peacefully deal with deteriorating situations before they become major threats to our security. Fully exploring and developing that array of preventive instruments will require the contributions of many people working over the equivalent of many lifetimes, and is far beyond the scope of this book.

What we can say here though, without the slightest equivocation, is that the way to start is by beginning to seriously address the three big evils that face us all together in today's world, once more: degrading, abject, hopeless poverty; the destruction of the environment, our "common wealth"; and, the degradation of human society and the proliferation of violence.

Even the most desperate, the most unfortunate and the poorest of the world's population would be given hope by seeing serious,

It seems to me, however, that we depart from a situation in which *the public already owns the key means of modern mass communication*, which are the television airwaves. Through the Federal Communications Commission in the U.S., current legislation allows the government – supposedly acting in the public interest – to license private operators to use specific television frequencies for commercial purposes. These licenses, the relatively limited number of frequencies available for television broadcasting, and the ability of private operator networks to acquire control of broadcast frequencies in a large majority of metropolitan areas, create quasi-monopolies which, in turn, permit these broadcaster networks to charge exorbitant amounts of money for access to the national airwaves and the mass national market that can be reached through them.

This is what makes it necessary, under the current system, for political parties and candidates to have a huge amount of money at their disposal, as, without it, there is no way for them to present and "sell" their message to a majority of voters.

That is just exactly how the two major political parties in the United States and their powerful corporate backers like it.

Since they control who gets the money they also control who gets to participate in the process. By controlling the field, they obviously also control who wins (or, at least, who does not win).

So television and the big money it commands for access to its all but monopolized mass market, has essentially become the "gatekeeper" who determines who is eligible to participate in the U.S. "democratic" electoral process and who is not. Thus, a new system has been surreptitiously set up that substitutes for but performs the same basic function as the property requirements for voting that were originally set up during the early part of our political history.

Far better than giving television broadcasters a huge amount of public tax money to finance political campaigns, let us simply legislate the recovery of a portion of the frequency spectrum and of a portion of the broadcast day for legitimate public purposes, including elections and other matters of urgent public concern. After all, let me re-emphasize, *the airwaves already belong to us!* Why in the world should we as taxpayers have to pay the networks huge amounts of money simply to have access to a medium that – by law and by common sense – is already ours?

The details will, of course, have to be studied and worked out carefully. But the broad outlines of a workable system are not too difficult to conceive. So many prime time hours during the broadcast week would, during electoral periods, be reserved to provide candidates with an opportunity to directly address the public with their messages[10]. Any candidate for public office with a certain minimum number of certified supporters, yet to be determined, would be able to address the pertinent voting public during the same amount of time as every other candidate running for that position, *at no cost either to the candidate or to the public!*

The cost, minimal indeed in comparison with the value of the licenses and franchises granted by the public to television stations and networks, would simply be absorbed as part of the cost of being granted use rights, for commercial purposes, of limited public resources, which are the available television broadcast frequencies, during the rest of the broadcast year.

Simply making it possible for a broad range of new voices to be *heard*, for the first time in the history of the American democracy, will profoundly change the way our system works. The stranglehold of the current two-party duopoly over the system will be broken. The people will for the first time have the opportunity to *listen* to competing voices regarding the direction toward a better future, think about and analyze a wide range of options without the intermediation of pundits, and freely *decide*, for the first time in their lives, how and by whom they should be governed.

This is all-important, and will make it possible to change everything else that needs changing.

---

[10] an additional blessing of the proposed new system: we'll be able to listen to the candidates directly, instead of to the endlessly repetitive drone of the "talking heads" and "pundits" *telling us* what the candidates are saying and what we should think about it

Subsidiary, but also important, topics for political and electoral reform would include the elimination of the Electoral College and the elimination of winner-take-all voting rules that currently exist in some of our political processes.

The Electoral College is an anachronism that no longer serves any useful purpose. The only two times in U.S. history in which different outcomes have resulted between the national popular vote count and the count of electoral votes – the elections of George W. Bush in 2000 and of Rutherford B. Hayes in 1876 – produced massive loss of confidence in the transparency of our electoral system. There is no good reason to even contemplate having this happen again.

As to winner-take-all voting rules, these obviously obliterate the sentiments of the losing minority or minorities, even though these may constitute a large fraction of the voting population. Where indirect voting systems are still in place, as in the capture and distribution of convention delegates in party primaries, a proportional distribution is clearly far more representative and far less likely to produce anomalous results in the final aggregate tally.

Economic and Fiscal Reforms

Freedom depends upon responsibility (and, of course, *vice versa*).

Individual freedom – the free exercise of our civil rights – depends upon transparent and fair adjudication and enforcement of the law.

The existence of a free market depends, similarly, on the availability of information and opportunities for all, on existence of a level playing field, and on the equal and transparent enforcement of treaties, laws and regulations that govern our economic interactions.

*Stop the Ripoff!*

The first and most important thing that needs to be done in the economic arena is to begin once more to enforce existing treaties, laws and regulations that today only mimic the governance of our economic interactions.

To begin with, the critically important and dangerous issues pertaining to freeing up market access, in countries like China, Vietnam and India, to foreign exchange, and to the prohibition of policies that lead to artificially undervalued exchange rates.

As was mentioned above, these countries currently maintain official exchange rates which, according to sources like the World Bank and the U.S. Central Intelligence Agency, are undervalued, with respect to the U.S. dollar, by a factor of *four or five to one!* This means that, if their currencies were allowed to trade freely and were to reach levels approaching purchasing power parity (PPP), the prices of their exports, expressed in U.S. dollars, would be 4 to 5 times as high as they currently are[11]!

This is an enormous distortion which is only made possible by these countries' governments' use of mechanisms to intercept and sequester foreign exchange reserves that accrue from their huge trade surpluses, which they then keeps off the market, only making relatively limited amounts of foreign exchange available through controlled rationing to selected importers at artificially established prices. It is easy to see that if China, which currently holds about 1.5 trillion dollars in FX reserves[12], were to release any significant amount of these reserves onto the market – thereby increasing their supply – the price of U.S. dollars and other international currencies released would fall dramatically in relation to the Yuan, meaning that many more

---

[11] This discussion is based on figures published by the CIA for 2006. They have recently been significantly revised for 2007, indicating for example that in their estimation Chinese GDP at PPP *fell* from 10.21 trillion U.S. dollars in 2006 to only $7.04 trillion in 2007, while the official exchange rate was only allowed to appreciate by about 5%. One suspects that political pressure in the U.S. may explain at least part of such a large year-to-year adjustment in estimates, but it has no effect on our current argument. Even at the reduced estimate, the Yuan, according to the CIA is *still overvalued by a factor of about 2.5:1.0!* Differences between PPP and official exchange rates between the dollar and the Euro or yen, for example, are only in the range of 10-15%, instead of the 404% reported by the CIA for China in 2006.

[12] Equivalent to almost 20 months worth of imports of goods and services at their current levels, far more than needed to assure them enough international liquidity. Also equivalent to about 15% of the foreign debt of the United States, a considerable piece of leverage.

dollars would be required to buy the same Chinese goods and that export prices would go up commensurately.

International trade and financial regulatory organizations like the WTO and the IMF put a lot of emphasis on reducing tariff barriers as a means of encouraging the growth of international trade.  An import duty of, say, 50%, imposed on the import of some manufactured product would almost without question be considered excessively high, and the country trying to impose such a duty would be strongly pressured to reduce it substantially.

Such grotesquely and extremely undervalued exchange rates as are managed by China and India represent rates of *effective protection* many times higher than the piddling 50% tariff given as an example above.  Depending on which estimates you believe, current official exchange rates in China and India are equivalent to import tariffs in the range of 250-500%!  They also represent a massive implicit subsidy to the export sectors of their economies[13], which allows them to sell their goods to foreign buyers at prices that are several *times* cheaper than what they would be if they were no longer allowed to hoard foreign exchange and keep it from becoming freely available on their national market.

The Chinese break the rules of international trade in many other ways that have also been mentioned above:  by making loans from state-owned banks to state-owned export enterprises that are never repaid, they are effectively providing a direct government subsidy to that exporting company, allowing it to maintain below-market export prices; by not enforcing environmental protection standards and fair labor norms, their government makes possible further cost avoidance by their exporting companies, making them even more "competitive" internationally.

---

[13]  The subsidy could also be considered to accrue to consumers in Chinese export markets like the U.S., though it would be hard to understand the Chinese government's interest in subsidizing U.S. consumers.  A more realistic interpretation is that the Chinese government is subsidizing the cost of buying market share in export markets around the world, by allowing Chinese industry to massively undercut prices below what any competitor can meet.  The cost of this subsidy is implicitly financed by reduced standards of living – an indirect consumption tax – levied on its own citizens, especially its rural citizens.

The impact of all of this is that unbelievably cheap consumer goods are flooding markets all over the world in those countries that continue to play by the rules. This is destroying entire industries in the United States, and in Europe, and in all of our trading partner countries in Latin America and Africa that do not impose retaliatory tariffs or otherwise protect themselves from these historically unheard of predatory trade practices. The flood of artificially, unfairly and – I would say illegally – cheap consumer goods from China into the U.S. market has destroyed hundreds of thousands if not millions of well-paying manufacturing jobs, that it has only been possible to replace with low-paying, dead-end jobs in the personal and retail services sectors. We can see and feel it happening, all around us. And yet, it is allowed to continue. Why?

The largest international ripoff in history is allowed to continue simply because a great number of U.S. corporations are making a huge amount of money having their products produced under license in China at a ridiculously low cost, and selling these at high margins to gullible and unwitting U.S. consumers. An important amount of money made from this trade – which, I repeat, is illegal in terms of the trade rules that apply to any other nation – finds its way into the campaign funds of leading U.S. politicians in the executive and legislative branches, and, of course, a lot of it also goes into feeding the advertising revenues of the large media conglomerates who also are encouraged to turn a blind eye or, at best, to only distract attention from the real problem – China – by talking about such irrelevancies as the U.S. free trade agreement with the tiny Central American republics or Peru.

China is being allowed to get away with it because of the corruption of our economic and political system. India and Vietnam are being allowed to get away with it, so far, because they have not yet become significant exporters, at least in comparison with the Chinese juggernaut.

So, the first and most important thing that needs to be done to redress the economic and fiscal damage being done by our broken system is to oblige the Chinese government, our own transnational corporations, and our political "leaders" to play by the rules that already exist. They are all, as things now stand, criminally liable to the public, and they will destroy our futures unless they are brought into compliance soon.

Taking action to redress this perversion of the market system will rapidly restore competitiveness to manufacturing industries located in the West, and will make possible the restoration of higher-paid and more stable manufacturing jobs. This alone will go a long way towards restoring a measure of decency and justice to the workings of our economic system.

There are, however, many other issues that need to be considered and, fundamental among these are issues pertaining to the elimination of extreme poverty, clarifying and strengthening the conceptual and legal basis for private property rights and overhauling our grossly inequitable and complex tax system.

### Ending Poverty – Let's Just Do It!

There is nothing that can make me understand why, in a rich country like the United States, I should be condemned to being minimally educated, being socially alienated and even criminalized, being poor and having to see my own children condemned to the same life all over again, merely because of where and to which parents I was born. Similarly, I don't see why the mere fact of your having been born into a wealthy family that was able to give you a good education, good business connections and a head-start at the beginning gives you the right to look down your nose at me, and blame my laziness, dishonesty or incompetence for my miserable situation.

Let's face it: even our work habits, character traits, talents and intelligence are mostly a matter of inheritance and luck. There is no defensible reason anymore for there to be destitute and despairing people in a land of plenty[14]. What is more, there is no defensible reason anymore for there to be destitute and despairing people in a *world* of plenty. Poverty can and must be eliminated, within a generation at the most.

---

[14] It is estimated, on the basis of the World Bank's *World Development Indicators* that Gross World Income per capita (PPP basis) is about $9,700. World income, equally distributed, would therefore provide all families of 5 people with a current income of about $48,500 in 2008. That is plenty, by anybody's estimation, to provide for a decent and dignified living for all. The wealth exists. What still is lacking is the will.

Again, most of the work of eliminating poverty in our country and in the world can be done by eliminating current distortions and inequities in our system that keep there from being a level playing field for all of us. But, there may also be more direct measures needed.

For example, we have already talked about education and health services. Making sure that all children have access to a good education and to good health services – also provided through the public education system if necessary – will go a long way to ensuring that the poverty gap closes within a generation. But, what of the home environment and of the lessons of despair and degeneration that extreme poverty can bring to children there? What can be done to make sure that all families can also afford at least the basics, and to keep themselves together as families that provide a nurturing home to all children?

One of the most creative thinkers on this subject in the last century, widely regarded as and in many ways truly an arch-conservative, was University of Chicago Economics Professor and Nobel laureate Milton Friedman. In his 1962 classic *Capitalism and Freedom*, he proposed a simple, cost-effective system for implementing a guaranteed minimum income for all families, based on what he called a "negative income tax". Basically, what he suggested was using the already existing income tax collection apparatus to also assign a certain amount of funds to all taxpayers and their families, thus replacing – with a simple, easily administered and low-cost system – all other bureaucratic, high-cost welfare programs such as Food Stamps and Aid to Families with Dependent Children. Those reporting less than a minimum amount of earned income and owing no tax would actually receive cash distributions from the IRS to help them attain a desired minimum level of income, while those with higher incomes and a positive tax obligation could credit their allowance towards their tax payment.

Actual field trials of a guaranteed minimum income maintenance system involving almost 9 thousand families, including both test groups and control groups, were conducted in the U.S. between

1968 and 1978.[15] Each family in the test group was given a guaranteed income by the government whether anyone in the family worked or not.  There were no strings attached.  The control group got no special treatment[16].

Results were surprisingly positive:

- few in the test group actually quit their jobs
- some cut the number of hours worked, by 7% in one group, by 6% in two others, and only by 1% in the fourth
- many in the test group used their guaranteed income to look for extra training or find better jobs
- 25% of the test group eventually (but within the relatively short lives of the trials, see footnote below) earned enough so they weren't any longer eligible for the minimum income

Evaluators concluded that cash assistance programs would not cause a massive withdrawal of workers from the labor force, as many had feared.  When combined with assistance in job search and access to public service jobs, it was determined that these programs would actually result in increased work effort.  These were the consistent results gathered from the careful observation and documentation of the behavior of almost 9,000 families in four study groups including different ethnicities and both rural and urban settings from Seattle to New Jersey and North Carolina.

Although legislative proposals to establish a Negative Income Tax in the United States were considered by Congress in the early 1970s, the times were turbulent and not yet right and these efforts were unsuccessful.  Another example of the triumph of *stasis* over change, and an unfortunate failure of imagination.

Imagine the payoff to our present society from the improved health, education and productivity of children who were raised in

---

[15] 1) Urban areas in New Jersey and Pennsylvania from 1968-1972, involving 1,300 families; 2) Rural areas in Iowa and North Carolina, 1969-1973, 800 families; 3) Gary, Indiana, 1971-74, 1,800 families; and, 4) Seattle and Denver, 1970-78, 4,800 families.
[16] Information on these experiments comes from a manuscript referenced on *Wikipedia* by Allan Sheahen, author of "Guaranteed Income: The Right to Economic Security," published in 1983.

families that were able to continue holding their heads high and maintain their dignity as a result of these programs.  Imagine the human tragedies and social costs that would have been avoided by saving even a small number of these children from drugs and delinquency.

It could have been done.  It can be done now, at relatively low cost, and with incalculable economic and social returns for our countries in the future.

Leave no child behind.

Leave no family behind.

Leave *nobody* behind!

What a world it is that could be coming if we just get started now!

### Redefine Property Rights and Responsibilities

What is "property" anyway?  What does it mean for a mortal human being to "own" a motor car?  A fine painting?  A good book?  A Labrador retriever?  A six-hundred acre farm?

What does it mean for a corporate "person" to own a steel plant, a coal mine or a 20,000 acre patch of forest?  A pharmaceutical patent?

These are highly-nuanced and complex questions.  There have been many definitions of property throughout the course of history and across societies and their legal systems, and they have always varied according to the type of property in question.  It is *not* the same thing to own a book as to own a dog or a piece of land.

In general, the idea of property can be conceived of as a bundle of rights and responsibilities.  It is *not* a relationship between people and things, but rather a relationship – subject to discussion and historical evolution and change – *between people,* with regard to things.

Traditionally, four major attributes of property rights are espoused:

1.  control of the *use* of the property
2.  right to any *benefit* of the property
3.  right to *transfer* or sell the property
4.  right to *exclude others* from the property

While the discussion is usually restricted to the rights conferred by property, in law and in practice and in all cases, property rights are restricted and also impose responsibilities on the owner of the property. All of which is sustained, in different ways from place to place, by social compact enacted into law to achieve a social purpose.

For example, I can "own" a vehicle and ownership gives me privileges in its use, compared to other people. But I can only use it, or allow it to be used, if the car is registered in my state of residence, if I have paid for and have affixed license tags in the designated manner, within certain limits as to speed, and on roads and highways or some few other locations where driving a vehicle is allowed. I have the right to benefit from the car, perhaps using it as a taxi, but only if additional licensing and other conditions are met. I can sell the car, but, again, only if certain regulatory requirements are met and if I pay whatever taxes may be due. Even the right to exclude others from the property is restricted in some cases, as for example, my obligation to allow a tow truck operator to take control of the vehicle and forcibly remove it if it is left in a prohibited area.

What of a book? I can "own" it in the sense of putting it where I want, reading it when I want, learning from it, selling it and keeping others from reading my copy if I so desire. But I can't copy or electronically reproduce it – supposedly in whole or in part – without permission from the copyright owner (for a certain number of decades until it enters into the "public domain"). I can't misquote it without exposure to possible penalty.

I can "use" my dog (whatever that means), but not "abuse" it. I can work my farm, and sell the crops that it produces, but I can not willfully or even negligently allow the soil to be eroded or the aquifers to become poisoned.

Corporations – which derive their rights from their recognition by the state as legal "persons" – are also restricted by law and by practice from the totally free and unrestricted exercise of their property rights.

Their use, usufruct, transfer and exclusion rights are all strictly limited by law, though perhaps in different ways, whether we are talking about a steel mill, a coal mine or a forest.

The legal definition of our rights and responsibilities with respect to both real and intangible "things" is what is meant by "property rights". Property rights are and have always been nothing more than a social construct, crafted in consonance with the beliefs and traditions of a particular society, and intended to facilitate the functioning of that society in accordance with some conception of the public good. There is nothing sacred about them, any more than any other kind of legislation that confers rights and imposes responsibilities on us for the public good.

It would be useful, in my view, to have a public debate about the kinds of property that we think, as a people, that it would be useful to define, and about the specific rights and responsibilities we think we should confer upon the holders of private property.

For example, a distinction between property that is "produced" by human effort and that which is received from nature is an important starting categorization. I would think that my rights in a chair that I have made or in a chemical formula that I have "discovered" would have to be somehow fundamentally different from my rights over a herd of reindeer or a deposit of gypsum that I happened to find.

Within the category of "produced" property, we could further distinguish between "improvements", "manufactures" and "intellectual" property. One would also perhaps want to include "labor" as a form of produced property that "belongs to us".

Our rights over produced property should obviously be very strong. But for our effort, or the effort we have "purchased" from others, these kinds of property would not exist at all. Because we as a society all benefit from each other's individual efforts to produce useful forms of property, our legislation should – and in fact does – confer very broad and strongly protected rights in this category.

Our "rights" over property that we have all – as the current tenants of this planet Earth – received from nature are a totally different matter, however. Here, the existence or non-existence of the "property" in question is totally independent of any of our individual efforts. This property – or more precisely, the responsible use of this property during our lifetimes – is a "gift" received by us from nature. No one person, individual or corporate, has any *a priori* claims on such property, and it is strictly a social question as to what rights should be conferred to individuals and corporations over the use of such property, and under what conditions, so as to maximize the value of our gift to ourselves and to future generations who also share in it. In fact, it is more useful to think in terms of "use rights" instead of "property rights" when it comes to these categories of property.

The gifts we all have received from nature, the common natural "resources" of the planet Earth, can be classified into "renewable" and "non-renewable". Renewable natural resources – forests, watersheds and soil, for example – can in principle under proper and responsible management be used forever without being depleted either as to quantity or quality.

Non-renewable natural resources – coal or tungsten deposits, for example – will sooner or later be exhausted and will have to be replaced by something else, probably more costly, in order to produce the benefits that we currently derive from their use.

Since these are fundamentally different types of property and since the laws and regulations that create rights and responsibilities over property clearly and directly affect the way that it is used, it is clearly in our interest, from a public policy perspective, to legally differentiate among them. In fact, much already existing legislation already does so, but without the clarity, consistency and coherence needed for the basic underlying principles to emerge.

Without going into any great depth, the basic principles that I would propose for consideration are that:

- to the extent that doing so does no harm to others, we should be free to use and to consume *produced property* pretty much at will and as we wish
- the use of *renewable natural resources* should carry with it the responsibility and legal obligation to bear the costs that will ensure that the resource is not depleted or diminished, either in quantity or quality; if this is not technologically possible, and the use of some critically important renewable resource necessarily implies some degree of degradation, then this should be kept to a minimum, and, through the imposition of user fees, corporations given access to the destructive use of such resources, and we as consumers ultimately responsible, should pay the cost of compensating future generations in some manner and/or of generating some kind of an equivalent replacement if that is possible
- the use of *non-renewable natural resources* should always and unequivocally require the payment of user fees – to discourage excessive consumption to the extent possible – and to compensate society for the costs of developing replacements or going without the benefits that we currently enjoy through the depletion of our gifts.

This simple framework can go a great distance in clarifying what we mean by "property rights" in a free, democratic and humane society, and make it possible to build the right incentives and disincentives into our framework of laws and regulations for the future.

### *Bring the World's Common Resources under Control*

Finally, and we will return briefly to this subject in our discussion of a new international regulatory framework to enforce the protection of environmental and labor standards universally, it will also be necessary to define and legislate property rights and responsibilities for resources that are now considered "common", such as the atmosphere, our oceans and the wildlife that lives within them. Unless properly regulated by a legitimate, competent and democratic international body, "common" property is nobody's property, and it is treated as though it had no more value to anyone than the short-term profits that can be gained by the most ruthless and shortsighted exploitation imaginable.

*Reform Our Tax System*

It is widely recognized that the U.S. tax code is overly complex, cumbersome and costly to enforce, and ultimately inequitable due to the relative ease, in comparison with wage earners, with which wealthy individuals and corporations are able to formulate effective strategies for tax avoidance.

Personal and corporate income taxes have the additional defect of discouraging saving and creating disincentives for growth. Proponents of the so-called Fair Tax and its variants recommend the replacement of most current federal taxes with a national retail sales tax, or, alternatively, with a national value-added tax. Both alternatives would provide a relatively simple and transparent alternative mechanism for generating the revenues needed to support the expenditures of the Federal Government, and, if combined with a tax rebate up to the poverty level as proposed, would avoid further harm to the economic position of the poor.

The Fair Tax "prebate" goes part of the way towards implementing a guaranteed minimum income system, as was suggested above, and could be expanded modestly[17] to also accomplish the broader goal.

The current federal income tax is slightly progressive in its incidence among households with different levels of income. That is to say, wealthier households pay a larger proportion of their total incomes than do lower income households. In fact, households in the lower half of the income distribution in the U.S. pay no or very little personal income tax.

---

[17] It is hard to find good or believable numbers on poverty in the United States. But, to use something like the official figures, there are approximately 15 million households containing about 25 million people below the federal poverty threshold, and the average "gap" per household of 2 is on the order of $7,000 annually. Fully making up this gap through a guaranteed minimum income program would cost about $135 billion annually, a figure that is clearly within our means.

However, because of the proliferation of tax loopholes that allow the rich to legally avoid the payment of taxes at maximum statutory rates, the personal income tax is also in fact gradually becoming "flatter" in the U.S., with an ever-increasing share of the personal income tax burden being shifted to the middle- and upper-middle income classes.

Payroll taxes, and in particular the Social Security tax, are quite regressive, with a relatively low level of income established as a ceiling for their application. Taken together, the personal income tax and the Social Security tax, which would both be replaced by a national sales or value-added tax under proposals currently circulating among some economists and legislators, are already quite flat or perhaps even slightly regressive. The potential impact of a change towards a consumption-based taxation system needs to be studied further, but it is not evident at the moment that such a change would lead to greater or lesser regressivity in the incidence of all federal taxes paid by individuals. Furthermore, the degree of regressivity or progressivity of a consumption tax-based system can be controlled by exempting certain classes of basic consumption goods and/or by adjusting the amount of rebates built into to the system to exempt lower-income families from any tax burden and, indeed, to make sure that they are able to enjoy a guaranteed minimum level of income.

A consumption-based system for the taxation of individuals will, very plausibly, be much more difficult for more powerful wealthy groups to distort through the gradual introduction of loopholes and tax havens, and will stop penalizing savings and investment, as is currently the case with the personal income tax.

Simplicity, ease, transparency and economy in enforcement, and the removal of disincentives to socially-beneficial behavior have much to recommend them.

In addition to the single consumption tax imposed on individuals and households, but in a similar spirit, this author would also propose developing a "user tax" to be levied on corporations that, in our name and for our benefit, "consume" non-renewable natural resources or are not able to fully mitigate the degradation of renewable natural resources used in their production processes. It should not be too difficult to establish usage rates for key resources by the U.S. corporate sector and to determine what would be a revenue-neutral level of user taxation that would fully replace the current corporate income tax (assuming that revenue neutrality is an important feature, which is not <u>necessarily</u> the case).  The proposed corporate natural resource user tax would have the strongly positive effect of encouraging corporations to conserve resources and spur the development of resource-conserving technologies.

Introduction of a simple system of "user taxes" to be paid by corporations in a manner comparable to the consumption taxes paid by individuals would also have the benefit of spreading the tax burden between individuals and corporations – allowing lower tax rates to be imposed in each sphere while maintaining any given overall level of revenues.

### Reduce Military Spending

Altogether, military-related expenses of the U.S. Federal Government totaled approximately $626.1 billion in 2007, including approximately $170 billion allocated to fund the wars in Afghanistan and Iraq.  That is about the same amount as the total GDP of the Netherlands or Australia, and more than the total GDP of 167 countries whose incomes are estimated by the World Bank; that is to say, a larger amount than the total GDP of all but 16 of the richest countries in the world, including the United States itself.  It is by far the largest amount spent on arms and armies by any country in the world, and absorbs about half of the discretionary spending in the U.S. Federal Budget.

Is there anything wrong with that?

Is there any way that wiser policies could help us save a substantial portion of this expenditure?

Yes.  Obviously.

We're on the wrong track.  We're not going to pacify the world, or remake it in our image, at the point of a gun.

Let's first try to minimize the threats we face by other far more cost-effective and productive means, even while maintaining the ability to protect ourselves fully 99.9% of the time.  Or, let's make it 99.999%, or whatever the appropriate level of security might be.  The point is, it is impossible to make all of us 100% safe all of the time, and we will make ourselves crazy, and spend ourselves into oblivion, if we try.

While, we should begin a gradual withdrawal from both Afghanistan and Iraq as soon as possible, it will be impossible to do so precipitously.  Therefore, while we must make it a priority medium-term goal to significantly reduce our spending on our military forces, it will unavoidably be necessary to embark on other programs first – programs that will enhance our security by stabilizing dangerous regions of the world through political and economic cooperation – then assess our progress, second. Only then can we begin a judicious long-term reduction in force, third.  But the process must start now.

Environmental and Labor Protections

*Uniform Global Standards and Enforcement*

As alluded to above, one of the major problems with our current system in this regard is that countries and laws for the most part only have national jurisdiction – and effective or even semi-effective enforcement only exists at the national level – while modern multinational corporations operate on a truly global level, and can usually easily escape or circumvent the jurisdiction of any one country.  It is very easy, and profitable, for corporations to *shop* the world for the least restrictive, least costly environmental and labor protection regulations.  And so, of course, they do so.  And we end up consuming products that will ultimately kill us, or our neighbors in poorer countries, or our children's children.

This obviously has to stop, and the only way to make it stop is by establishing a new level of international cooperation to set uniform – or at least compatible[18] – standards for environmental and labor protection.  Enforcement, to be effective, inescapable

---

[18] It is perhaps valid to consider allowing slightly more lenient standards for poorer countries during a transition period in which other adjustments will need to take place.  For example, many poorer countries in the world still have a longer work-week than do richer countries, and it would bring hardship to workers to suddenly reduce the maximum legal number of work hours until hourly wages have a chance to adjust as well.

106

and uniform, should be conducted by some international body analogous to the WTO, with strong, uniform penalties imposed, with the support of the new body's member governments, for infractions committed anywhere in the world.

## Common Resources

Secondly, more effective international cooperation is required to control the exploitation of global resources – such as fisheries and some wildlife – that do not fall under the purview of any national government, or perhaps only under the purview of relatively weak governments needing assistance in preserving their natural resources base. Uncontrolled fishing, conducted under the inexorable pressures of the marketplace, is leading to the extinction of species after species. The overall catch is in decline, and increasingly contaminated by mercury or disease, just as we are discovering the crucially important health benefits of including fish in our diets.

Whales are being hunted to satisfy the depraved appetites of Japanese *epicures*, tortoise eggs continue to be offered to consumers looking for a sexual lift all along the coasts of Central America and Mexico, sea lions are slaughtered by the thousands by poor, ignorant fishermen in Peru who can't survive the competition, elephants, tigers and mountain gorilla are disappearing from the forests of Africa and Asia to make way for more hopeless human misery in ever more tragically degraded environments.

Etc., etc., etc. So, the creation of another WTO-like organization specifically to look after wild species all over the world that currently "belong to nobody" is urgently needed.

## Medical and Pharmaceutical Industries

Third, it is important to do something meaningful about the direction that is currently being taken by the medical profession and pharmaceutical industries around the world. Conducted within a market-driven framework, medicine and medical research are forced into a profit-maximizing mode that is ultimately biased towards the preservation of life under conditions of chronic illness, over the prevention or cure of disease. There is no, or very little, money in prevention or cure.

So it doesn't happen anymore. It is just as simple as that, and we are all paying the price, both monetarily and in a deteriorating quality-of-life.

The cost of keeping people sick-but-alive is exorbitant, and is breaking us as individuals as we reach old age, and as national healthcare systems as our entire populations get older. We must find ways to restore incentives and eliminate barriers in order to foster greater investment and effort by the medical profession and medical and pharmaceutical researchers in preventive and curative approaches to disease. This may involve direct action or subsidies by the public sector, greater international cooperation in research and development, licensing of practitioners of alternative forms of medicine, spurring the development of a global market in medical services, more effort to prevent the adverse effects of environmental toxins on public health, etc.

It is a large and complicated subject. However, the first step to finding good, practical solutions – which has not yet been taken by our governments – is to recognize that the system is not currently working as it should, and that there may well be more effective alternatives available. The second step, and this is where we as citizens come in, is for our governments to find the moral fortitude to resist the phenomenal pressures that can be brought to bear by vested interests in the current system – including specifically the trillion-dollar medical, pharmaceutical and health insurance industries.

It comes back to politics and the influence of money in controlling politics under the current, distorted, system. It can be fixed, and we can fix it within something like a decade – not more – if we get started now.

Restoring Security and Societal Integrity

The third great evil which we identified at the outset as being one of the integral if unintended by-products of the working of our system is social disintegration and violence. Closely related to the first great evil – our abandonment of billions to a destiny of abject poverty – social disintegration and violence must first be addressed at the systemic level, through clear and decisive action to eliminate the biases and destructive feedback loops that currently work to transmit and concentrate the impacts of negative exogenous shocks on the weakest among us, and which pit the poor against each other in a cold and cruel global price competition for the commercial favor of the rich. These mechanisms, which – I believe – developed almost "naturally", with little thought, premeditation or deliberate action on the part of anyone over the course of thousands of years of human history, are inherently unfair and unjust. Security and social cohesion can never coexist with injustice, and we must now choose one state or the other.

### Walking the Talk

So, the most important things that need to be done to enhance our security and make our societies more stable and cohesive are things that we have already talked about at length:

- make sure that the rules of international trade and competition are enforced equally and fairly
- make sure that regulations governing the protection of the environment and of workers' rights are enforced equally and fairly around the world
- clarify the conceptual basis of property rights and responsibilities, distinguishing between produced property and the gifts we have collectively received from nature; reform regulatory and fiscal policies accordingly
- establish new international institutions with authority to oversee the enforcement of environmental and labor protections, and to exercise jurisdiction over what are now considered to be global resources
- really leave no child behind
- move aggressively and fast to redress the most extreme inequities in incomes and the current inability of millions of families around the world to secure their basic human needs for survival and dignity

We can do all of this, if we decide to. And now that we know that we can do it, we must do it, or lose our own souls in the balance.

In addition, there are some other specifics that can also be identified, and recommended for further analysis and evaluation.

*Creating Effective International Security Organizations*

First among these is the strengthening and democratization of international organizations – currently existing or to be created – that are concerned with the maintenance of international security. The United Nations Security Council is one such organization. NATO is another. There are many others already existing, with different degrees of functionality and importance.

One of the key problems of such organizations at the present time is their governance. Either, as in the case of NATO and other regional security organizations set up under U.S. auspices, they are regionally-, ideologically- and militarily-based and therefore exclusive, or, as in the case of the United Nations and its affiliated organizations, are widely inclusive but riddled with inconsistencies and incoherencies in terms of the rules of their own governance.

For example, the United Nations General Assembly currently has 192 member states, and it is governed by the rule of one-state, one-vote. A great many of the states represented are very far from being democratically-governed themselves, and their votes are therefore only representative of their rulers and their interests, not of their populations. Secondly, one-state one-vote means that very small, weak countries have the same vote as large, strong countries. It has been estimated, for example, that countries representing only 8 percent of the world's population could, theoretically, organize a 2/3 supermajority in any vote of the General Assembly.

Giving credence or standing to the votes of the U.N. General Assembly would obviously be absurd and foolhardy under the circumstances and, in practice, this body is largely ignored by the leading world powers.

The U.N. Security Council is made up of five permanent members that each have veto power (China, France, Russia, the United Kingdom and the United States, in customary alphabetical order), and 10 temporary members who are selected to participate for two-year periods by the General Assembly. Even here, two of the five permanent members represent governments that are very far from being representative democracies and therefore can only be assumed to vote the interests of their ruling elites, while the other three members' electoral and government practices could also stand considerable improvement, as we have been discussing at length in prior sections. Usually, when an issue comes before the U.N. Security Council, one or more of the five permanent members has a vested interest in the outcome, and it has been difficult to develop a procedure that would lead to a different outcome than what would result anyway from the direct interplay among the great powers on the world stage.

So, what good does it do?

A good question without any clear answer for the moment. Presently existing international institutions either lack legitimacy and credibility, or lack effectiveness, or both.

It is really difficult to conceive of international bodies becoming effective in the maintenance of international security in the absence of fundamental changes in their membership and in the way they govern themselves. But, international security issues are obviously of such critical importance and so threatening to us all that we can not continue to leave them simply to the law of the largest gun.

One approach might be to build in two directions simultaneously. One, continue strengthening existing regional security organizations, with a primary focus on the settlement of disputes among regional members, and on organizing a common defense against possible attack by non-members. These organizations are heavily influenced by military considerations, and probably should continue to be so, although these can of course be broadened – as they already are by leading military thinkers – to include consideration of the military and security implications of economic, social and political policies within their regions.

A second, perhaps new or reformed international organization growing out of the thinking that originally established the League of Nations and the U.N., would be an organization of united *democratic* nations where the first order of business among members would be mutual assistance in the perfection of their own internal systems of governance. In the process of establishing the "U.D.N." the founding members would of course be challenged to define the minimum standards to be met by members for admission – a difficult but undoubtedly most productive and rewarding task in itself.  Next would come the task of developing fair and effective voting rules to govern the new organization.  This will call for a great deal of creativity, but something useful can be devised to allow the organization to establish its credibility, maintain the support of its members and function effectively.  Some kind of weighted voting scheme would probably need to be developed among the members of the U.D.N., perhaps using a mix of population weights, on the one hand, and weights based on some realistic measure of relative power on the other.  Perhaps the relative weights given to population and to "power" could be designed to alternate during successive periods in time, giving members an incentive to structure and maintain as balanced and equitable a system as possible.

However it is to be done, we need functioning and effective international organizations to deal with the manifold and proliferating problems that can only be dealt with through international cooperation.  The only other alternative is the hegemony of a new imperial power over the entire planet, and that is clearly not the way to peace and international security.

### Dealing with Drugs and the Criminal Drug Trade

A great deal of street violence, much corruption of public officials at all levels of government and huge financial and social costs originate from illegal drug production, distribution and consumption.  No one knows exactly how large the numbers are, but estimates of the street value of illegal drugs consumed yearly in the U.S. are on the order of $400 billion.  That is a huge amount of money, which goes to empower a major international criminal apparatus, which, in many settings around the world, has also established links and operational alliances with international terrorists.

In short, we are doing a lot of harm to ourselves – especially our young – and we are paying dearly for it into the bargain.

The time has come, in my opinion, to decriminalize private personal drug use under controlled conditions of production, distribution and consumption that can ensure that these substances do not continue to find their way into our junior high schools and the bodies of innocent children.

This is not because I would advocate the use of drugs of any kind. Although, like many of my generation, I experimented to a limited degree with marijuana and LSD back in the '60s and early '70s, nowadays I don't consume anything stronger than a good single-malt whiskey, and don't trust or like to take even an over-the-counter medicine unless I really can't avoid it. I personally would tell anyone: don't take anything you can't control, especially strong stimulants, narcotics or other psychoactive substances that can become strongly habit-forming.

Just say no to drugs.

A well-funded, serious and truthful public education program should accompany decriminalization as a way to change the public perception of drug usage – away from the image of daring, excitement and fun, to one of illness and drug dependency – and thus help young people avoid its allure and the temptation to try it "just once".

The main, compelling reason to decriminalize drug usage under controlled circumstances is to destroy – once and for all – the huge monetary value that is currently generated by this global criminal activity, and all the violence and corruption that also accompanies it. If drugs were provided in controlled, limited quantities to licensed adult drug users in a distribution system similar or identical to that currently being used for the distribution of prescription drugs, and if these were produced under license and public control, the illegal drug industry – like the illegal liquor industry after the end of Prohibition in the U.S. during the last century – could be destroyed overnight. Unlike liquor, which is a far less harmful but still dangerous substance, there could be no commercial or public use of drugs, only private use of limited amounts in their own homes by licensed adult users who would, in the course of obtaining their licenses, be exposed to good medical advice and given options to receive medical treatment instead. Any resale or transfer of these controlled substances to an unlicensed user, or the operation of motor vehicles or other dangerous equipment while under their influence, would be treated as criminal offenses just as is currently the case.

It is not easy to accept the fact that our children may be exposed to substances as harmful as these, and to the lifestyles that accompany their usage. However, that is the reality we face today. And, as things stand today, their exposure is being effected by criminals who operate without any legal or moral constraints, and whose only interest is in creating as much dependency as possible. The so-called "war on drugs" has been going on for at least four decades now – at the cost of trillions of dollars and millions of human lives destroyed. Far from having been won, it has become an industry unto itself, with more and more people around the world – from the fields and laboratories of Afghanistan, Colombia and Mexico to the squad cars and courtrooms of Los Angeles, to the schoolyards of suburban Washington, D.C. – finding that they have a vested interest in its continuation. The moral "rot" that sets in as a consequence of our failure to take effective action against such a clear and present danger – again, mostly because of the huge money that can be made by allowing it to continue – is as bad or worse as the damage done by the usage of these substances itself. This rot is one of the main elements that is destroying our societies in the U.S. and parts of Europe, and it has to be stopped, with intelligence instead of violence.

### Dealing with Public Pornography and Violence

Finally, there is another element of "rot" in our contemporary society which also does great harm to moral values and social cohesion. This is the deluge of pornography and violence that we are currently experiencing through every medium of public communication, whether that be the internet, cable, broadcast television, magazines or billboard advertising. Again, driven by an unbounded lust for money, pornography and mindless sadism teach our children that there are no real values worth upholding, and that it is Ok or even smart to use other people, *and themselves,* in any way imaginable so long as they can make some money. Our acquiescence as parents to having their lives and our lives inundated and corrupted in this manner only serves to confirm this sad reality in their young minds.

What are we teaching our children?  What kind of a world will they have to live in, where gang bangs and snuff movies become the stuff of everyday life?

The industries that produce pornography and images of depravity and violence make huge amounts of money and are served by the most highly-paid and skillful lawyers and lobbyists, who have somehow managed to convince us that any restriction on the public distribution of their ugly wares – of their version of "rot" – is an infringement of our own liberties!  How has that possibly come to pass?  We have to allow ourselves and our children to be subjected almost continually to the most sordid and degrading of images and ideas – in the name of liberty and free speech!  Why not let ourselves and our children be actually raped and butchered instead?  Why not simply let – as, little do we realize, is actually is happening as we speak – the virtual become real.

Let the virtual become real.  The virtual *is real*, and we have to bring it under control.

The pushers tell us that it is up to us as parents to control what our children are allowed to see.  They say that their sleazy products are only available in the midnight hour, when children should all be in bed anyway, and that they have provided special programmable chips and other controls that can limit their access to unsuitable materials.

It is all lies.  The stuff is everywhere, around the clock, and it is impossible to avoid.  Even kiddie comics have been infected, and the innocent fantasies of infants have become rotten.

Not only that, it is enticing.  That is why it sells.  Pornography and violence appeal to deep-seated, dark, reptilian instincts that are hard-wired into us and very difficult to control.  But control them we must, or society and civilization mean nothing at all.

How and where to draw the line between protecting ourselves and our children from the massive and mortally dangerous mental and moral assault that we are now being subjected to by the craven pushers in our midst – and falling into the equally dangerous trap of allowing ourselves and the artistic products of our minds to be censored by dogmatic governmental or ecclesiastical authority – is a very difficult question to resolve exactly. But it is easy to see that, wherever that line may lie exactly, we are way, way beyond it now. It is time to take control again, and establish standards of public decency and morals that can help us and our children achieve a higher level of humanity.

Pornographers and pimps are the scum of the earth and sadists are deeply twisted and dangerous. Let's get them off the public airwaves and out of our children's and our own vulnerable and suggestible minds. Get back under that rock! There will be no more money for you in this society.

# A Summation: What Is Wrong With the World?
## And What You Can Do About It

**I.  What's Wrong?**
   a.  All human beings have a right to a decent life; the current system doesn't allow that to happen; it is designed by and for the powerful to perpetuate their privileges without regard and/or at the expense of the misery of others; it is unsustainable and is only leading to ever more cruelty and violence, at every level of existence; non-violent means have to be mobilized and used uncompromisingly to permanently change the current system.
   b.  We are destroying the ecological and resource base of the planet, to indulge wasteful levels of consumption by the few, while the costs will ultimately be borne by the many; this also is unsustainable.
   c.  We have lost our moral compass; life now is about self-indulgence rather than about discovery, sharing and spiritual development; without internal direction to lead us towards a fair, healthy and happy planet, order is maintained increasingly through propaganda and police repression. Nihilism and corruption deeply undermine this schema, making it also ultimately unsustainable.

**II.  What is the Good?**
   a.  The good is what allows us to maximize the value and enjoyment of our lives through discovery and development of our essential spiritual natures and our dedication to bringing about a fair, healthy and happy planet that will remain so for future generations.
   b.  The good is sharing what we have, what we know and what we see.
   c.  The good is helping to relieve the suffering of others, and taking pleasure in the enjoyment of others.

**III.  What You Can Do to Change the World**
   a.  Believe that you can.
   b.  Believe that you must.
   c.  Find others who believe what you do.
   d.  Understand what is really going on around you.
   e.  Resist the forces that seek to stifle or distort your understanding and your beliefs.
   f.  Resist nihilism and corruption.
   g.  Speak the truth.
   h.  Share your beliefs and your resources in order to fulfil your dreams.
   i.  Never give up.
   j.  Live for Life. Live for your ancestors, for your country, for your friends and your little green planet. Live for God and your humanity. Live for a fair, healthy and happy planet for you and your children.

# V.  Responding to Our Immediate Crises

This final chapter focuses on a few current issues and crises that are perceived to be so threatening and of such importance that they must be addressed by any author writing on contemporary affairs, or risk being accused of deliberately avoiding the most difficult and contentious questions of the day.

These are, first of all:

What to Do About Al Qaeda?

This is certainly a most difficult and contentious question, and my thoughts are offered in all humility.  As in the analysis of any other difficult situation calling for action, however, one should begin to address it by clearly defining our own priority objectives:  in my mind, the clear, overriding priority objective of what we do about Al Qaeda is to ensure that we are not subjected to another attack such as took place on September 11, 2001.

Guarding against such an attack requires the best intelligence possible, a high level of security at likely targets, and the ability to respond quickly to disable any group that is identified as being engaged in suspicious activity against us.  This seems to be pretty much what the U.S. and allied security forces have been doing since 2001, with what appears to be, overall, a pretty high level of success, despite the tragedies in Madrid, London and Bali.  Things could have been a lot worse during the last six plus years since 9/11 and in general our intelligence and security forces are to be congratulated.  Their work needs to be continued, and continually reenergized and reinforced until the day finally comes that the threat has been significantly reduced.

Good intelligence, good security and the ability to take quick preventive action are *not* the same, however, as the invasion and occupation of Iraq.  Although the Bush Administration claims that, by instigating the war in Iraq we have shifted the battleground against terrorists to a distant shore, what is in fact far more likely is that we have fueled a massive expansion in anti-Western feeling and terrorist insurgency in the Middle East, and have provided a training ground for terrorists that will fill their ranks for years or even decades to come. Al Qaeda is already present in almost every continent of the globe – from the Philippines, Indonesia and Malaysia through to South Asia, the Middle East, East and North Africa, Europe and America – and, unless we somehow manage to kill them all, which seems unlikely, it would appear that even in the event of the total pacification of Iraq, the best that can happen is that, like the mujahideen veterans of the anti-Soviet campaign in Afghanistan, former insurgents in this war will go underground and disperse to points all over the world, where it will become even more difficult to extricate them. Going into Iraq on the pretext of removing phantom WMD, seems to have left us in between two very hard places – let the war in Iraq continue as a way to attract and give focus to global Islamic insurgency, or bring the war in Iraq to a close and allow that insurgency to metastasize with renewed virulence around the globe. It's hard to see how anything good will come from the current path we've been put on by the Bush-Cheney Administration.

Be that as it may, it will obviously have to be dealt with for some time to come, in the first instance by redoubling our intelligence, security and prevention efforts and capabilities, as described above.

Secondly, as has been learned from earlier conflicts, we have to face up to the huge challenge of finding good ways to "drain the swamp", and "win the hearts and minds" of the hundreds of millions of Muslims around the world who are not violent insurgents but who hold deeply-felt legitimate grievances against the policies and actions of the U.S. and other Western states in their countries.

We must begin by recognizing that, in very large measure, "the swamp" is our very own creation, and that it is in fact our *responsibility* to drain it and siphon off the many alligators that we have allowed to prosper in it at the expense of the ordinary citizens who occupy these lands.

The Islamic countries are largely backwards politically and socially, in part because of the rigidity of their own traditions but also, in large measure, because the West has deliberately kept it so. Except for Turkey, there is not, so far as I can recall, a single functioning democracy in *any* of the over two-dozen predominantly Muslim countries of the world, most of which only achieved independence from Western colonial powers in the period between the end of WW II until the 1970s. Democracy obviously had not flourished under colonial rule and, by and large, despite our own rhetoric and traditions, it has not been recognized as being in the Western interest to support the growth of democracy in these areas since. As in Iran and Egypt, the West has actively intervened to suppress fledgling indigenous attempts to institute democratic, secular governments, and, elsewhere, we have armed and supported archaic and corrupt monarchical and dictatorial regimes that have been willing to serve as the guarantors of Western energy supplies during our long struggles against the spread of communism during the Cold War.

And now, it will be extremely difficult to be rid of them.

However carefully, in light of the ubiquitously looming threat of violent political upheaval in so many countries of the Muslim world, we must nevertheless now let it be known that we will stand firmly behind a rapid transition to democratic rule in these regions, and that we will no longer grant legitimacy to governments that are instituted or maintained by force against the will of their people. The West as a whole, with the assistance of our allies among the Asian democracies, must also be prepared to actively support the transition to democracy in the region, knowing that it will be a long and costly process, with a coherent program of technical assistance, market access and development financing where that is also needed.

Let us be under no illusions. This situation is a mess. But we got ourselves into it and we must now accept that it will take a long time to get out of it. And a strategy based on the continued repression of the forces of change in the region will not work, and would in the end be far more costly for us – in resources, human lives and our own prospects for a better future.

So let us start doing right, where, up until now, it has seemed easier to keep on doing wrong. Let us use – and be seen to be using – our resources and our power to bring about prosperity and peace, where we are now making war. Let us insure that a free, democratic and peaceful Palestine reemerges to grow and prosper alongside a free, democratic and peaceful Israel. If we do this – while guarding always against extremist attack – I believe that Al Qaeda will gradually wither away as an instigator of violence against us. If we don't, I am afraid that we are all in for a world of hurt for decades to still to come.

## What to Do About Illegal Immigration?

Like international terrorism, illegal immigration and undocumented workers are big problems in most of the richer countries of the West, from the United States to France, Spain, Germany and Great Britain. Given their individual geography and histories, the problem nevertheless has its unique features and characteristics in each country.

The U.S. has a special problem with illegal immigrants from Mexico and Central America; France with Algerians and Tunisians, Spain with Moroccans, Germany with Turks and Eastern Europeans, Great Britain with South Asians. Recognizing these differences, it is possible to develop more effective programs in each recipient country to address the problems at their source: insufficiency of jobs and opportunities to make a living in the countries of origin. While recipient country governments are well aware of this, there has not yet been enough public education on the subject to develop a strong constituency supporting an expansion of economic and technical assistance to poor countries that are themselves facing the disastrous consequences of a process which is leading to their loss of large numbers of their better-trained and more ambitious working age population.

The first response to the problem of illegal immigration to the U.S. and elsewhere should therefore be a concentrated effort to motivate job creation and more rapid economic development in the specific countries of origin. These are not many, and the cost need not be extremely high.

Secondly, as has been done in some cases, the recipient countries need to officially recognize that a great many of their currently undocumented workers are in fact needed to sustain the competitive and efficient functioning of their own economies. In many Western countries, the indigenous population is aging rapidly and educational levels are such that few citizens are interested in filling jobs requiring low-skilled physical work. Despite the numbers of undocumented or "illegal" workers who are presumed to be active within their economies, Western countries have in general had very low rates of unemployment during the last decade, indicating that their labor markets are indeed demanding the work that is being done by both legal and illegal workers. And, despite the rhetoric and if only through their consumption activities and their obligatory payment of sales taxes, local governments in fact do collect a substantial amount of revenue from undocumented workers, which at least partially offsets the cost of public services that they may receive.

More importantly, if undocumented workers are able to work in economies that are at full employment, then clearly their work is producing a multiplier effect in relation to the overall level of activity – income generation, consumption and investment – that is taking place, and thus is also contributing to public revenue collections *throughout* the economy in question. A more balanced public discourse on the subject of "illegal" immigration and undocumented workers would perhaps help to educate the populace more fully as to the positive contributions these workers are making to the recipient economies, and would perhaps contribute to the generation of support for more balanced and objective, more moderate, approaches to bringing the situation under control.

In the United States, for example, politicians of all stripes who are currently campaigning for the presidency are pretty much boxed in to having to say that they will initiate programs to force all 12 million or so "illegal" residents to have to go back to their countries of origin, and "go to the back of the line" in terms of their ability to apply for legal re-entry. Some candidates are saying that this can be accomplished by some kind of massive national police dragnet in only a matter of a few months, while others talk about giving people a reasonable amount of time to put their family affairs in order before having to leave.

They are all talking nonsense. Assuming the numbers are anywhere near being accurate, can any of you imagine what would be involved in identifying, rounding up and forcibly expelling 12 million people from within our borders? The backlash and disruption would be unimaginable.

123

Besides which, the reality is that we *need* the majority of these people. They are performing useful work, and they do a good job.

So that perhaps a more moderate and rational approach would be to establish a process whereby recipient countries can document and "legalize" undocumented workers without necessarily conferring either permanent residency or citizenship status. This would essentially imply the development and implementation of a "guest worker" program, where people are allowed to reside and work legally for a given number of years, then have to return to their country of origin, unless they have transitioned to permanent residency or citizenship in the meantime.

This is not the same as "amnesty". In principle, all workers who are not admitted to residency or citizenship would have to return to their country of origin after a specified period, but in an orderly and dignified manner. Many such workers in fact desire to return to their countries of origin, once they have been able to earn enough money to provide honestly for themselves and their families.

The necessary law enforcement effort to expel illegal aliens who are not gainfully employed or who are, even worse, engaged in criminal activities could then be concentrated on where the serious problems really are, at much less cost, much less disruption, and much less violation of human dignity. With a well-organized guest worker program in place, it should be relatively easy to secure the cooperation of employers in implementing it, with no penalties for either employers or employees who cooperate, and strong penalties against employers who continue to hire outside of the program.

Having laid out a strategy and a framework for a program that rationally addresses both the problems that are leading to the expulsion of people from their countries of origin and the problem of the large number of undocumented workers who are already here, then a coherent immigration policy could also focus on stopping the future flow of illegal immigrants across our borders and through our ports. Reasonable build-up of the numbers and capabilities of our border patrol and immigration agents, and reasonable investments in physical facilities and technology should make it possible to do this much-reduced and less draconian job at much lower cost than what the politicians and the media are mostly talking about today.

## What to Do About China?

The relationship between the U.S. and other Western nations and the People's Republic of China is fraught with danger for all parties.

On the one hand, China can not be allowed to continue abusing its status as a state-controlled economy to maintain a hugely undervalued exchange rate and all sorts of proscribed practices that currently give it an insurmountable competitive edge in world trade and are gutting manufacturing industries throughout the West.

On the other hand, the West is already so bought into and exposed to the currently distorted terms of Chinese trade, and it owes so much to China, that it can not afford to press too hard for overly rapid reform.

The Bush Administration and the current Congress have engaged in some theatrics, designed primarily for U.S. public consumption, that have led to a couple of token adjustments to the Yuan-dollar exchange rate – an appreciation of 1.0% in 2005, followed by 2.7% in 2006 and 4.5% in 2007 – all laughable in comparison with the 250-400% by which institutions like the CIA and the World Bank estimate the Yuan to be undervalued.

However, can anybody imagine the impact on consumer prices and economic activity in the U.S. if China were suddenly to make a realistic adjustment to its currency value? We could be plunged overnight into a stagflation that would make that of the 1970s seem like a joke.

What if, in addition, China were to decide to stop buying U.S. Treasury bonds to refinance the current over $1 trillion in U.S. bonds that it currently holds? What if, beyond that even, China were able to persuade – in light of their common long-term mutual interests – some of our other important Asian creditors, who are also key trading and investment partners with China, to also slow down their investments in U.S. Government securities? We could, almost overnight, be plunged into a Depression that would make that of the 1930's seem like a joke.

In all likelihood, such a sequence of events would also sooner or later plunge us into a new world war that would make WW II seem like a joke.

The U.S. and other Western governments have been almost criminally irresponsible in allowing our economies to become so extremely vulnerable to the vagaries of Chinese policies, determined by a small group of dictators, mostly old men with long memories who dream about restoring Chinese power to its millennial status before the rape of China by the West in the 18th and 19th centuries. They do not have to worry about electoral politics or other internal pressures in the short-term, and, if necessary, are quite prepared to have their populations tighten their belts for a decade or so while the process of finally and decisively destroying the economic and military power of the West unfolds.

What to do?

Wake up to the extreme danger that we are in, to begin with.

And then?

Dance, and keep on dancing, in the hopes that we can gradually redress the extreme dependency and exposure that our corrupt leaders and their corporate sponsors have allowed to develop in the 1990s and early this century.

I would guess that this dance would begin by letting the Chinese know that we have finally heard the music, that we understand what is going on, and that we are determined to redress the situation peacefully if possible. We would need to convince them that we really do intend peaceful coexistence and cooperation – if possible – and that the current situation is untenable for either party. A gradual process of adjustment must begin to take place if either one of us is to survive.

Something like what Ronald Reagan was, somewhat miraculously, able to achieve in staring down the Soviet grizzly bear, and convincing Soviet leadership that their only viable options consisted in *glasnost* and *perestroika* and the relaxation of military tensions between us.

A dangerous process at best, but unavoidable at this point. And it has to begin by letting everyone know that we finally understand what is going on, and that we are serious about bringing it to an end.

More exact and reliable calculations of purchasing-power parities need to be developed and used as a guide for the estimation of effective protection rates in force in countries around the world, for their reduction and for the formulation and implementation of countervailing trade policies, not only with regard to China, but with other currently less-formidable exporters who also control their currency values, such as India and Vietnam.

We must also insist on the gradual but deliberate and steady enforcement of stronger environmental and labor protections in countries, including China, that are currently not complying with international norms.

As a strong side condition to the gradual peaceful adjustment of current trade distortions and imbalances, it is imperative that the Chinese government also accept the fact that a free market economy can only exist within the context of political freedom and democracy domestically, and of full insertion and commitment to a transparent and equitable trading system internationally.

In order for that imperative to become clear to the Chinese, and for it to become acceptable as a basis for policy, it will also be necessary for us to be willing to show our own commitment to a gradually reduced militarism in the conduct of our foreign policies and its implicit threats to Chinese security.

In the meantime, we urgently need to get ourselves back into a position where we can respond quickly to protect ourselves both economically and militarily in the event that the Chinese leadership decides – as it could at any moment – to adopt a more aggressive and perhaps apocalyptic posture towards the resolution of our differences.

Shall we dance?

# Epilogue:  Musings on the Meaning of Freedom

If you and I are both free then we can both do pretty much whatever we want, so long as we don't hurt the other in a direct, clear and intentional manner.

I can't tell you what to do, and you can't tell me what to do.

I am not responsible for you, and you are not responsible for me.  We are each responsible only for ourselves.

Neither one of us is responsible for the world around us.

No one is responsible for the world around us, and what goes on in the world around us basically "can't be helped" and can't be changed, by me or by you.  Even if I do hurt you, if it is not evidently clear how I am hurting you, if it is indirect (say through what I do to the environment or to our common culture), then there is nothing you can do about it.  I am "free".

We both eventually, then, become irresponsible "free riders", ultimately driven only by cynicism, sensation, hedonism, self-gratification and self-interest.  "Everybody else is doing it", so it must be alright.  At least, it can't be helped.  There's nothing you can do about it.  Whatever "it" might be.

Until not very long ago, we were all responsible to "God" and God's Ministers walked about among us, telling us what God wanted us to do and not to do, "for our own good".  Those of us who did not follow God's rules (as we were all "free" not to do) were recognized by all of us to be "sinners", and to be deserving of punishment, both in this life and the next.  For most people, this seemed to be enough to keep us from doing too much harm to one another on a day-to-day basis except in ways (slavery, exploitation, war) that apparently "God" thought were Ok, necessary from time to time, or good tests of our characters.  But, more generally, on a daily basis people lived thinking of themselves as individually responsible to God, and all of us as responsible to one another, through God's "Church".

Nowadays, God has moved further away from us, to a less comprehensible place, and his Ministers no longer seem to be so fit to tell any of us what to do.  So none of us is responsible any more, and the world is suffering greatly because of it.

What if, though, what if each of us *was* individually responsible for each other and for the state of the world we live in?  Let's say, by individual epiphany and sacred promise, without need for recourse to a higher authority. What would that then mean for "freedom"?

I guess that what that would then mean, first of all, is that we would all of us have to start by coming together and deciding on what, minimally, is "good" for the world, "good" for children and other living things, "good" for people, "good" for our culture, etc. That's obviously not such an easy thing, but if we don't get overly ambitious too soon, if we concentrate on just the bare minimum that we can all agree with, maybe it could be done.

Second, we would need to confirm our individual commitments to "the good" and accept our individual responsibility for sustaining it. That means, also, "freely" accepting those limitations on our individual "freedom" that are necessary – first to keep us from going off the deep end, and, second, to justify our occasional interference in the freedoms of others who, by some due process, are deemed to be in danger of going off the deep end.

Could we accept such limitations and still be "free"?

In principle, yes, if we can all agree on the conditions that will bind us.

The practical problem then becomes how to bring about a broad enough consensus to involve a large enough number of people to actually make a difference in the state and condition of the world, without having to water down our standards on what is "good" beyond the point of insignificance. The tradeoff is between the quality and rigor of standards to be upheld, and the breadth of consensus to be achieved.

In principle, it could be possible to achieve a broad consensus at a very high standard, but probably not now. It would take education and persuasion, over an extended period of time. So the task now is, find the set of positive but not impossible standards that will allow us to achieve a "dominant coalition", and begin to act according to those standards, freely accepting our individual responsibilities towards one another and the world at some broad level. With time, practice and the experience of success, we may then set our sights on "freely" bringing about an ever higher good. We shall then have become "free", in order to be "good".

## Phillip W. Rourk
*Biographical Note*

Phil Rourk is an economist and international consultant with over 30 years' experience advising international organizations, government agencies and private businesses in more than two dozen countries around the world.

Included among these assignments were: directing a sector-wide planning exercise for the energy sector of Thailand in which a team of over 35 engineers and economists from around the world helped to identify, evaluate and sequence about 50 billion dollars of investment in natural gas extraction, petroleum refining, and electricity generation; serving as chief economic advisor to the Ecuadorian Minister of Finance in the late 1980s in negotiating a Stand-By Agreement with the International Monetary Fund and renegotiating payment schedules for over $5 billion in debt owed international commercial banks, and about $2 billion in debts owed OECD governments; preparation of feasibility analyses that ultimately led to the first private infrastructure development investment undertaken in the former Panama Canal Zone, the $111 million construction of the first stage of the Manzanillo International Terminal, a load-center container port that is today the largest in Latin America; conceptualization and execution of a program designed to assist small- and medium-sized companies in El Salvador become successful exporters, increasing their total exports by more than $80 million in less than 5 years.

He was born in Nicaragua to a U.S. Foreign Service family. At age four, his family was transferred to Beirut, Lebanon, and he began school at a French school for Lebanese children, learning French and Arabic in the process. He continued to study in a French school at their next posting, to Rotterdam, Holland. Next, they were assigned to Bogotá, Colombia, where he attended the Abraham Lincoln School, his first English-language school, and then to Washington, D.C. where his father served a tour at the U.S. Department of State.

His university studies were in economics, architecture, philosophy and fine arts, and he received a B.A. in Philosophy from the American University in Washington, D.C. in 1970. He then returned to Nicaragua, where he managed a family-owned coffee farm near Matagalpa for several years until returning to the U.S. to attend graduate school in Economics at the University of Maryland.

On obtaining an M.A. in Economics in 1977, he joined the international consulting staff of Robert R. Nathan Associates, Inc., where he remained for 12 years, ending his service there as International Vice President, responsible for managing the firm's consulting business in Asia and Latin America. In 1989, he left Nathan Associates to form a new independent economics and finance consulting company, The Americas Group, Inc. This company was succeeded by the AG International Consulting Corporation in 1991 and the Washington International Finance Corporation in 2001, with both of the latter small companies remaining active to the present day. In 2003, he founded the FairShare Foundation, Inc. a 201.c.3 tax-exempt organization designed to channel contributions from individuals and corporations towards productive social investments in disadvantaged communities in the U.S. and developing countries around the world.

His three daughters live and work in the Washington, D.C. metropolitan area.

# APPENDIX –

## Unpublished Political Essays

## and

## Correspondence

# Must We Perpetuate the Madness?

To paraphrase (and greatly oversimplify) James Welles[19], probably our greatest academic authority on the subject, stupidity is willful persistence in behaviors that are maladaptive at best, counterproductive, self-defeating and self-destructive at their worst.

Here we go again.

What, exactly, do we expect to accomplish by, one more time, bravely assembling our military might and launching cruise missiles into the night? Against "terrorism"?

I guess we think that we can kill "terrorism". That, after all, is what militaries and cruise missiles are for.

Undoubtedly, we can kill terrorists, and undoubtedly we will. In greater or smaller numbers, along with who knows how many "collateral" kills.

But we can't kill "terrorism". Not with cruise missiles.

Is terrorism madness? Is it mindless, heartless evil, incarnate? If it is, then we must fight it with more appropriate weapons. Madness can not be eradicated by madness, nor evil with yet more evil. As both Jesus and Mohamed taught, we must deal with madness with understanding, compassion and intelligence. And only goodness and love can defeat evil, beginning with the evil in ourselves, then the evil that resides in others. It's not as though we haven't been told about these things before.

Is there, perhaps, some rational basis for the terrorists' actions? Are there, at least in their own minds, grievances serious enough to justify their cruel and desperate acts? If there are, then we clearly must seek to identify and understand those grievances. If they are real, we must correct them. We must also *do* justice if we seek justice.

If the grievances are not real, then we must show that they are not. In any case, we must show that they do not justify – in a world that genuinely seeks justice – the kinds of violent acts perpetrated against us on September 11. Any more than they can justify the kinds of violent reprisals we are now once again embarked upon ourselves.

---

[19]   James F. Welles, Ph.D., author of <u>Understanding Stupidity</u>, 1986, and <u>The Story of Stupidity</u>, 1988, both available from Mount Pleasant Press, Orient, New York.

Is terrorism a combination of rationality *and* madness, of righteous grievances *and* evil?

Yes.  And it can't be killed with bombs.

Let's get over (after how many thousands of years, how many millions of dead?) this stupid, maladaptive and self-destructive behavior.  Let's get real.  Let's grow up.  Let's *stop* the madness, on their side *and* our own.

Phil Rourk
October 7, 2001

October 9, 2002

Editor
The Washington Post

## Saddam – Indict, Arrest, and Try

May I suggest an alternative approach to the one-and-a-half pronged approach (pretend to push for an effective inspection regime while preparing for all-out war) currently being followed with respect to the obviously real threat posed by the continuation of Saddam Hussein in power in Iraq?

While the International Criminal Court may or may not be the appropriate venue, it is evident, given its repudiation of that court earlier this year, that the Bush Administration is currently in no position to call upon it at this time. Alternatively, I would suggest that the U.N. Security Council, under U.S. leadership, convene a special "grand jury" to consider criminal charges against Saddam Hussein and his closest associates in crime. Based on the evidence that is common knowledge to all, it is almost certain that a multicount indictment could be rapidly brought. Whereupon, the Security Council could proceed to call for the immediate surrender of those indicted, or order and organize the forces that would be necessary for their arrest. This would be a far different thing than waging war against the country of Iraq to achieve the same purpose.

Whereas war will almost certainly produce large scale destruction of innocent life and property in a poor country that is itself already the victim of Hussein's tyrannical regime, following internationally-sanctioned indictment it is highly likely that those around him will recognize their own self-interest in his isolation, betrayal and possible capture. Under these circumstances, with the good intelligence produced by such defections, it should be possible to effect his capture with a precisely-targeted attack aimed specifically against his most proximate defenses. Once we know where he is, and once we have been legitimately empowered to effect his capture, control of the air deployed in combination with a concentrated and heavily-armed strike force of five thousand or so can almost certainly isolate and capture or kill him in short order – sparing the lives and future hopes of hundreds of thousands of innocent Iraqis that will otherwise be jeopardized. Not to mention the very real risk to our own soldiers, our citizens and/or the citizens of Israel if a conventional military attack triggers his use of weapons of mass destruction.

Most important, our country will have acted as it should as the primary defender of the rule of law in the conduct of international affairs, rather than as some new pretender to imperial power that somehow conceives itself to be above any but its own corrupted concept of law. The power of such a precedent and its implications for the future course of human history are enormous.  At stake is nothing less than the soul of America, and the fulfillment of America's true historical destiny.

Phil Rourk
Bethesda, MD

## We Need More Flexible Instruments

Everybody knows or thinks they know that Saddam Hussein is a tyrant. It also seems certain that he has lusted over weapons of mass destruction, and that he has devoted massive amounts of the meager treasure of a poor country to developing them and their means of delivery. We don't know exactly what he has now, but there is no doubt that his possession of such weapons constitutes a dire threat – difficult to gauge in terms of its immediacy – on the security of Israel, the balance-of-power in the Middle East, control over Middle Eastern petroleum resources and, ultimately, the security of the United States and the "West".

Does that mean that war is justified?

Modern war means all out aerial and ground assault on the military forces, economic infrastructure and civilian populations of "enemy" countries, and I suggest that, despite the threat posed by the criminal Hussein, war against Iraq – its conscript soldiers, means of economic survival and urban civilian population – is far, far from having been justified.

In the first Gulf War, to "liberate" Kuwait, Kuwaiti and Saudi oil, we killed about 100 thousand Iraqi soldiers and who knows how many civilians, ultimately only to preserve the *status quo ante*, in both its favorable and unfavorable aspects from the US perspective.

This time, how many innocents are we prepared to kill? It could be a very large number, and many of them could be our own. If that's the course we follow, we need to be very clear about it, and prepared to accept the responsibility it implies.

Still, we can't just sit here and do nothing while Saddam pursues the bomb.

Right. We need to develop other instruments to allow us – and the responsible global community of world citizens generally – to respond to threats of this kind without having to pull the trigger on all out war and the carnage that it inflicts on people who are themselves *victims! Not* "enemies".

Alternatives? The best one I can think of is some kind of international juridical process. If Saddam Hussein – or any other rogue "leader" around the world – is truly suspected of having committed crimes against humanity, then he/she should be formally indicted by some standing (International Criminal Court) or special *ad hoc* international body. With that kind of formal review of evidence, due process and considered judgment, an arrest order can be issued, and a *targeted*, not frontal, assault can legitimately be launched, specifically with the objective of capturing the indicted felon so that he/she can be brought to trial. That can be accomplished with far smaller destructiveness and loss of life than recourse to all out war. Indeed, in the best of cases, it can be hoped that the erstwhile colleagues of duly indicted international felons will themselves do the work of capture and delivery, sparing the lives of many, including their own.

In the worst of cases, it may be impossible to penetrate the tyrant's defenses in order to conduct an arrest. In those, absolutely worst of cases, I could envision a duly constituted international body, upon consideration of sufficiently compelling evidence of past crimes and imminent threats, issuing a warrant for capture "dead or alive" – a "termination with extreme prejudice" sanctioned by at least some semblance of legality, and a mechanism which, unlike war, is subject to control, refinement and improvement over time.

There is, absolutely, no justification for all out war against any people other than self-defense against an imminent and credible threat. That is not currently the case in Iraq. The wanton slaughter of American and Iraqi soldiers, Iraqi and probably Israeli civilians, and the destruction of the few capital assets of an already poor country is a price too high to pay simply to guard against a vague unlikely risk of attack, or to effect "regime change" with who knows what to follow.

We need to start thinking very carefully about, and to urgently get to work developing, more flexible political instruments. War is usually not the answer, certainly not now.

Phillip W. Rourk
October 19, 2002

## You Idiots in Government and Big Media: Will You Please Listen?[20]

My obviously simple-minded argument is as follows:

1.　　There is no credible evidence that Saddam Hussein presently *has* any militarily significant weapons of mass destruction, or the means to deliver them.
2.　　Even if he did, he is not stupid or suicidal enough to use them in a first strike against *anybody*. He knows full well what kind of hell would rain down upon him immediately if he did.

Therefore,

3.　　There is no compelling reason to launch an invasion of Iraq at the present time. Other approaches (such as have been described and completely ignored in prior submissions by this author to the Washington Post) need to be developed and given a chance to work before we pull the trigger on war, and all the slaughter of innocents and stupid destructiveness that war implies.

But, let me grant all of you idiots the *whole* of my simple-minded argument, and the *totality* of your worst suspicions. Let's say that Saddam *does* possess WMD and also the willingness to use them "preemptively". If that were the case, it has to be a given, an *a priori certainty*, that he will *also* use them if attacked. *We*, in our inimitable high-tech cowboy style, begin by launching a first wave of aerial attacks, and *he* immediately launches nuclear-laden SCUDS (or a network of sleepers positioned so as to spread smallpox, or whatever other diabolical engine of destruction we can imagine him to currently possess) against Tel Aviv, or New York, or Chicago. Thousands, maybe tens of thousands, are killed immediately or fatally infected.

*What then*?

Well, I guess that any *freedom-loving* American would understand that, at that point, there would be *no option* but to pull out all stops and *immediately* obliterate Baghdad and its general vicinity with our own nuclear attack.

*Wow!* And you guys call that military *strategy*? Astute, farsighted, foreign *policy*? If you all really believe what you claim to believe, is *that* where you think we should be heading?

---

[20] Submitted to the editors of the *Washington Post*, and also forwarded to then Assistant Secretary of State John S. Wolf and Secretary of State Colin Powell.

Absolutely idiotic. Maybe, just maybe, the counter argument I represent is worth examining a little more closely. There *are* other means available to all but the most gonadally-challenged, to all those able and willing to think just a little more creatively and responsibly, before we launch this global catastrophe in the making. Please see – and publish prominently – my earlier submissions to this newspaper. Thank you in advance.

Phil Rourk
Rockville, MD
October 22, 2002

# 95% of What Jeremiah has been Saying is Right!

Let's not unfairly pillory Barack Obama for the other 5% that we can all recognize as the rantings of an egomaniac.

- Were our white forefathers responsible for enslaving Africans and slaughtering native Americans? Yes.
- Are African-Americans as a group still disadvantaged relative to European-Americans as an enduring legacy of slavery? Yes.
- Did lynchings and church bombings continue to prevail in the Southern United States throughout the '60s? Did I personally drink out of a water fountain marked "whites only" in our capital's Union Station in the early 1960's? Yes.
- Were Medgar Evers and Martin Luther King murdered because of their race and the stands they took to end racial discrimination in the United States? Yes.
- Is a large amount of residual anger among African-Americans understandable and justified in light of their historical treatment in the United States of America? Yes.
- Has the present Bush Administration, and numerous other U.S. administrations prior, blatantly lied to the American public on subjects as sensitive as the true evidence supporting decisions of war and peace, and U.S. policies on torture and political assassination? Absolutely.

It may not be pleasant to look at and we may not like to hear it, but all of that is true. If Jeremiah Wright spoke for years on these subjects, and Barack Obama listened, they were both right, to be incensed in the first place, and to want to do something about, in the second. It is important that the truth be told, no matter how unpleasant, and we must face up to all of this if we are, in fact, going to finally move beyond it. That, I think, is what Barack Obama is all about. Facing up to where we have all come from. Getting all that behind us, finally. And moving forward, together, to what all of us want to see America become.

Based on his own recent confirmations and admissions, over the last 20 years the Rev. Wright has also said a number of things worthy only of a loon.

- Are his claims regarding the alleged U.S. government role in spreading HIV/AIDS among African-Americans and engineering the terrorist attacks of 9/11 dead wrong? Certainly.

But were these statements representative of the core of his message? No. Almost certainly, 90-95% of his time in the pulpit – and of Barack Obama's time in the pews – was spent on other subjects like those enumerated above.

Unpleasant? Yes. But true.

- Did the Rev. Jeremiah Wright, who served over 6 years voluntary service in the U.S. armed forces during the Vietnam era and three decades as pastor of an African-American church, do a lot of good for his Chicago parishioners over the last 20 years? Probably.

Should we believe that Senator Obama's membership in the Rev. Wright's church over a period of years, and his recognition of the true and valuable elements of Wright's service to the African-American community, means that he agrees with the loony 5% of Wright's rantings over the years? Absolutely not.

Let's face up to our reality, past and present. The Wright gambit is a transparent attempt by Senator Obama's opponents to force him into a defensive position, to force him off his strong message of change. It should not be allowed to work, and should certainly no longer be presented by the media in such an unreflexive and biased manner.

Time to move on, America.

Unanswered Letter to Bill O'Reilly, Fox News, May 2008

# DO WE REALLY NEED A PRESIDENT?

In 1996, our current President was elected with 47 million popular votes out of a voting age population of 197 million. Only 49 percent of the electorate (i.e., 96 million) participated in the election, and, of this, 49 percent voted for the Democratic candidate, 41% for the Republican, and about 10% for the Reform Party and other independents. So in fact, at the end of the day only 24% of the electorate actually voted *for* the President! An astonishing 75% majority either voted against the "winning" candidate or, by declining to participate, against the electoral process itself.

Again in this year's unfolding elect-o-drama, it appears that at best we will have a new President with about 24% of the electorate behind him. At worst, we will have someone with an even smaller percentage of the electorate, and a *minority* of the popular vote! It's unbelievable!

Consider how deep the irony, and how serious the problem, when you realize that Nicaraguan Sandinista Daniel Ortega, in *losing* his 1996 presidential bid had a larger proportion of the Nicaraguan electorate vote in his favor than did the "winner" of either of the most recent American presidential elections. Proportionately, fewer people support the man who occupies the most powerful office in the world than supported the *loser* in one of the poorest, least democratic, worst-educated and most backward countries of the world.

Not once in the last 50 years has an American President been elected by a majority of the electorate. Is it any wonder that so many Americans feel more and more alienated from their government when it is presided over by a person whose constituency seldom includes more than twenty to thirty percent of us? How much more so when we consider how rigged the "two"-party system has become anyway, when our choices among possible national leaders are ever more blatantly manipulated at the outset by the Tweedledums and Tweedledees of our political system, our two national political machines, big media, and their big money controllers?

It has sadly become a commonplace to observe that the system is no longer working.

The important questions are, however:

- Are there structural or design features of our two-hundred year old system that are causing it to fail? I.e., is there something we can fix? Or, as some would have us believe, is our system failing just because of us and our cultural or educational or civic shortcomings as a people?

And,

- If there in fact are structural problems that we can identify and that are keeping the system from delivering on the Founders' promise of self-government, is it still possible for us to gather the will to respond effectively and fix them? What could be done and how could we do it?

Let's begin by examining the role of the Presidency and of a President in our Republican form of government.

Unlike the case of parliamentary democracies where, especially in a multi-party context, national governments are often a seemingly "accidental" by-product of an electoral process that can be easily dominated by local and sectoral interests, the existence of a Presidency and the separate election of a President allows for the electorate to focus its attention and express its views on common, national issues separately from its narrower local and/or sectoral concerns. The Presidency is a uniquely national institution, and the President is at least supposed to "represent us all".

Apart from the undesirable effects of the two-party duopoly system on the partisan way in which it sometimes works, by and large the U.S. Congress functions as it was intended to. It is a highly representative body, and, as the innumerable minorities and special interest groups that make up our richly diverse country become better organized and sophisticated, it seems to become more and more representative as time goes by. Sometimes very slowly and laboriously – sometimes not – the Congress does its work. Laws are passed – on everything from civil rights to taxation to the environment – through a process that generally allows for widely disparate opinions to be heard and to be considered. With notable exceptions like the tax code, by-and-large the Congress has legislated wisely and well, managing somehow through a fractious and highly competitive process to find the right balance between the public interest and protecting our freedoms and our daily business from unnecessary government intrusion.

Taken as a whole, the corpus of American law is a marvelous construction, unique in the history of the world and rightly an object of great pride for our legislators and for ourselves as a people.

Broadly, it would appear that the Judiciary Branch of our government is also working reasonably well. Issues are (eventually) adjudicated. Most citizens, most of the time are able to get a fair trial, and due process is generally observed in both civil and criminal matters. The laws are enforced and individual rights are respected.

But, in the gaping schism that has developed between an institution which is by design supposed to "represent us all", and the reality of an electoral process that, for decades now, has produced Presidents who only represent a few, a major structural flaw in the Presidency is becoming increasingly evident. The institution, now embodied in the person of a single President, is simply not doing what it is supposed to do. And, because it is such a highly visible and symbolic institution – that is supposed to represent us all but really doesn't – many of us have lost faith in our system generally. Sadly – more than sadly, dangerously – many of us, especially among the young, have stopped participating in it at all.

Is the idea of a national institution "representing us all", or at least a large majority of us, at any given time simply an idealistic notion that should not, and perhaps was never intended to be, taken seriously? Is it possible for one person, a President, to embody that institution and effectively represent a large majority of the people?

The answer to these questions, both logically and historically, is "Yes".

There can and have been times, as when facing a great external threat for example, when a people, despite its possibly great diversity, does come together around an issue or a

146

cause and the leader that has arisen as its champion.  Franklin Delano Roosevelt and George Washington are examples.  There may be others.

Also, there are cases of societies that are either so homogeneous or (and?) so repressive as to "naturally" produce great uniformity of thought among their people, and the logical possibility of near unanimity in their support of a single individual as their leader.  One thinks, hypothetically, of Japan and 18th century Puritan New England as possible examples.  Maybe of Teddy Roosevelt as the embodiment of the swaggering WASP male-dominated, ethnocentric and nationalistic American culture of his times.

Modern democracies must – practically by definition – be able to cope with great diversity however, even as they strive to maintain cohesion and loyalty around the notion of a single nation state.  As diversity is recognized and protected for its intrinsic value, and as minorities proliferate around a myriad of ethnic, religious, economic, cultural and issue-charged polarities, it has become increasingly difficult if not impossible for any single individual to fulfill the purpose of the Presidency.

Except perhaps in a time of dire national emergency, it is no longer possible for any one man or any one woman to even come close to "representing us all".  And if it is impossible for any President to represent more than a small minority of us at any given time, is it wise to continue entrusting so much power to that institution?  Viewed from the perspective of the nowadays generally excluded (but diverse) majority, it would appear that the potential for harm that comes from granting so much power to someone we mostly disagree with probably exceeds by far the benefits of maintaining that single, national institution.

Aside from its symbolic and unifying leadership functions, the Presidency as an institution is derived from a number of historical and practical antecedents prevalent at the framing of our Constitution.  First, of course, is the historical and cultural legacy of monarchy.  The President is an obvious king-figure, and as everybody knows, but for the greatness of soul, humility and self-discipline of our first national leader, we may well have had an American king.

The Founders were undoubted geniuses of political innovation, but in this regard, that is in the design of the Presidency, mostly to the extent of doing away with hereditary features of the "monarch" and replacing that method of transmission of authority with a periodic electoral process.  Beyond that, it was widely thought – and this belief was strongly supported by tradition and the practicalities of governance at the time – that a single national executive was necessary to the functioning of a national government. Distances were large, communications were poor, Congress met only infrequently, etc. and so it was necessary to entrust a single person with the power and the authority to conduct the nation's business, taking decisions and acting day-to-day *in lieu* of the people and their legislative representatives.

And it was right to think so, at the time.  In order for a nation to be governed, you needed either a Monarch or a President.  Two hundred some odd years later, this is no longer the case.

You have only to look at the European parliamentary democracies, where presidents don't exist or play a purely ceremonial role, and where the power or the very existence of monarchies have disappeared over the years, to see that directly-elected national chief executives such as presidents are not *needed* in order for a government to function effectively.

So that we are entitled to conclude that here in the United States where we are faced with a Presidency that is no longer functioning properly, we *could* well decide to eliminate it entirely, and things could well improve in terms of the quality of governance in this country, as well as in terms of inclusion, citizen participation and self-identification with government.

Alternatively, recognizing the very great value – not afforded to the people in parliamentary democracies -- of having a *national* institution to focus our electoral attention on *national* issues separately from sectional issues, we might begin to think about ways of *reforming* the Presidency rather than eliminating it entirely, principally to make it a more truly representative institution and to safeguard us from the possible excesses of a single individual who does not represent a majority of us.

One way in which this might be done would be to amend the Constitution so as to invest the powers of the Presidency, not in a single individual but in a group of individuals organized into a Presidential or Executive Council.

For example, the top three popular vote-getters in a national "presidential" election would be elected to the Council, where their individual choices in reaching the decisions of the Council would each be weighted in proportion to the number of popular votes they each obtained in the presidential election.

One could even go so far as to elect whatever minimum number of candidates it takes to reach a minimum threshold level of representation of the electorate, say fifty-one or sixty percent, perhaps even two-thirds. Such a Presidency would:

- by construction, always represent a majority of the American people, while respecting, reflecting and benefiting from its diversity
- allow for decisions to be made by a majority consensus among leaders legitimately claiming to represent a majority of the electorate
- protect the populace by restraining possible extreme actions or abuses by any individual member of the Council
- preserve the ability to make rapid executive decisions on critical issues where a clear consensus exists among the members of the Council and, presumably, the people they represent

The one potentially grave defect of a Presidential Council is the theoretical possibility of gridlock, a situation in which it is not possible to establish a majority consensus among the members on some issue of critical importance to the nation.

Presumably, this could only happen in rare instances where no one member held a majority of the Council votes (because of a very divided popular election) and where opinions held by the Council members on the issue in question were so strong and so different from one another as to preclude the possibility of negotiating a consensus. In such a case, the division among the members of the Council on the gridlocked issue would presumably also necessarily reflect a deep division on this issue among the populace at large.

One, perhaps the only democratic, way of dealing with such a situation would be to require that either a popular referendum or a presidential election be held to resolve the issue. Certainly, the present way of entrusting a decision on such grave and contentious issues to a single person not representing a majority of us is not democratic.

Faced with gridlock on a serious issue within the Presidential Council, the Constitution could empower any member of the Council to call for a national referendum or election. The Constitution might require that the referendum or election be held quickly, say within a month or even a week, during which time a highly focused campaign would be conducted by advocates of all contending views, and the people would be given the opportunity to decide this presumably grave and contentious issue based on awareness of all the options for action or inaction and their consequences.

If a referendum, then the sitting Presidential Council would receive its clear mandate on the contentious issue and otherwise continue its business as usual. Or, if an election, a new Council presumably reflecting a new alignment of decision-making authority among its (possibly new) members would be installed.

Modern communications technology, applied to higher purposes than at present, makes this form of self-government a perfectly viable alternative for the American democracy as it enters into the 3d millennium. Given less restrictive and less costly access to television for political leaders than at present (which can be accomplished simply by legislatively mandating public access to the public airwaves for high public purposes), a wide diversity of messages can be rapidly transmitted to reach virtually all of the electorate in a very short period of time. The people can respond almost instantly, if need be, by incorporating the Internet or simply touch-tone telephone or "smart-card" technology into our voter registration and vote collection systems. Certainly, the integrity and security of individual votes can easily be protected, just as they are millions of times a day for credit card transactions.

We have it in our power, for the first time in human history, to literally govern ourselves.

It is a thrilling prospect for humanity, and a challenge to conventional thinking from which we as Americans should not shrink.

By adapting the structure of our governmental system to an evolved cultural and political reality, the relatively simple Constitutional reforms envisioned here would:

- allow a much greater diversity of voices to be heard in our national public discourse, reflecting the diversity of our people that is a key source of our nation's greatness
- foster the development and participation of a larger number of parties in our political system, without compromising the ability of our government to develop an operational consensus and reach executive decisions quickly
- greatly foster citizen participation in our nation's political life, through their involvement with parties that genuinely represent their particular interests and

concerns, and through their direct participation in decision-making on important national issues

- restore the relevance, functionality and integrity of the Presidency as our executive branch of government, while rescuing this most highly symbolic of our national institutions for the people, away from the clutches of political ego-trippers, pollsters, cynics, manipulators, spinmeisters and others of their all-too familiar ilk.

As to whether or not we as a people still have the will to make the changes that we need, only we as a people can decide.

Phil Rourk
August 2000

Phil Rourk is an economist and international consultant based in the Washington, D.C. area.

---

Historical Notes:

An Executive Council of Three was advanced and strongly advocated as an alternative to a single "President" by several of the delegates (Benjamin Franklin, Edmund Randolph and George Mason, among others) to the Constitutional Convention held in Philadelphia in 1787. The proposal was ultimately defeated, but only by a 7 to 4 vote. Randolph and Mason felt so strongly about the dangers of an excessive concentration of power in the single President that they ultimately refused to sign the Constitution at all, despite occupying leadership positions in their delegation and having contributed greatly to the deliberations of the Convention.

In 1878, after the failed presidencies of Andrew Johnson and Ulysses S. Grant, and widely-perceived fraud in the 1876 election of Rutherford B. Hayes (who lost the popular vote to Democrat Samuel Tilden but was able to engineer an electoral college majority of one), a proposal to amend the Constitution to replace the President with a three-man Executive Council was moved in the U.S. Congress, though of course not approved at that time.

While the impetus behind an Executive Council in the U.S. was historically largely based on mistrust of excessive concentration of power in one man and the abuses this could lead to, the Swiss – in their efforts to build a strong democratic nation from a multi-lingual (German, French, Italian and Romansch), multi-ethnic and -religious collection of local groups – directly faced the issue of incorporating and benefiting from diversity in the structure of their executive branch. The Swiss Constitution of 1848, in other ways greatly influenced by the American Constitution, consequently provided for a Federal Council made up of seven members, serving for a four-year period during which the largely ceremonial "presidency" of the Council rotates among the members on a yearly basis. The inclusiveness of the Swiss Federal Council, currently composed of members affiliated with four different political parties, has been an important factor contributing to the consolidation of a free, democratic, prosperous, stable, and humane society that nourishes and protects the diversity of its heritage and culture.

# AGENDA FOR U.S. CONSTITUTIONAL REFORM

Phillip W. Rourk
August 2000[21]

## I. Excerpts from the Current Constitution

### PREAMBLE

*We the People of the United States, in Order to form a more perfect Union, establish Justice, insure domestic Tranquility, provide for the common defense, promote the general Welfare, and secure the Blessings of Liberty to ourselves and our Posterity, do ordain and establish this Constitution for the United States of America.*

### Article V.
### AMENDING THE CONSTITUTION

*The Congress, whenever two-thirds of both Houses shall deem it necessary, shall propose Amendments to this Constitution, or, on the Application of the Legislatures of two thirds of the several States, shall call a Convention for proposing Amendments, which, in either Case, shall be valid to all intents and Purposes, as part of this Constitution, when ratified by the Legislatures of three-fourths of the several States, or by Conventions in three-fourths thereof, as the one or the other Mode of Ratification may be proposed by the Congress. Provided that no Amendment which may be made prior to the Year One thousand eight hundred and eight shall in any Manner affect the first and the fourth Clauses in the Ninth Section of the first Article, and that no State, without its Consent, shall be deprived of its equal Suffrage in the Senate.*

## II. Proposed Amendments to the Current Constitution

### THE LEGISLATIVE BRANCH

1. Establish term limits for members of both Houses of Congress, of up to four two-year terms for members of the House of Representatives and up to two six-year terms for Senators.

---

[21] Amended October 4, 2007 to eliminate an education element in the proposed weighted voting system.

# THE EXECUTIVE BRANCH

1.  Replace the single President with a Presidential or Executive Council, comprised of the smallest odd number of candidates for the Office of Member of the Executive Council as shall together represent at least two-thirds of citizens of the U.S. who are eligible voters as of the date on which the presidential election is held.

2.  Establish that each Member of the Executive Council shall have a vote in the decisions of the Council that is weighted in proportion to his or her share in the total popular vote. Decisions of the Council shall be determined by simple majority.

3.  Whenever a majority can not be established in voting for any question before the Council, any Member may require that the question be put to the People to be decided by popular Referendum to be held within a period of not more than 30 days. To have binding effect upon the Council, the outcome of the Referendum must be determined by a majority of the votes of at least two-thirds of the electorate.

4.  If the outcome of a Popular Referendum is mooted by the lack of a majority or a deficiency in the extent of voter participation, the Executive Council shall be obliged, within a period of not more than 30 days, to call a General Election to elect new members of the Executive Council. Sitting members of the Council may run for re-election, up to a maximum of three successive terms. Former members may again become eligible for election after a minimum of one full term out of office.

5.  Barring terms having extraordinary mid-term elections, the normal term of Members of the Executive Council shall be three years.

6.  Any U.S. citizen otherwise meeting the requirements of the current Constitution for the Office of President may be elected to serve on the Executive Council, without regard for the location of his or her birth.

# THE ELECTORAL PROCESS

1.  All elective offices of the Executive, Legislative and Judicial Branches of the Federal Government shall be filled by the direct popular vote of the electorate residing in the jurisdiction in question or outside of the U.S.A. on Election Day. The current Electoral College is abolished.

2.  The electorate is composed of all U.S. citizens who are at least eighteen years of age on Election Day, and who are neither incarcerated nor legally judged to be mentally incompetent as of that date.

3.     Each voter shall cast a number of votes that is determined by the voter's age (years of life experience). I.e.,

    a.   Each voter shall be entitled to cast 1 vote on having celebrated his or her eighteenth birthday.

    b.   Each voter shall be entitled to cast 1 additional vote for every ten years of life beyond the age of twenty, up to a maximum of 5 votes on reaching the age of sixty.

4.     All broadcasters licensed by the U.S. government to utilize the public airwaves shall be required to provide the equivalent of one-hour daily, in total, without charge, jointly to all qualified candidates for public office during the month prior to an Election. This time will be made available between the hours of seven and ten p.m. on weekdays, and/or between nine a.m. and ten p.m. Saturdays and Sundays. Equal distribution of this time (thirty "prime time" hours during the month prior to the election) individually among all qualified candidates shall be adjudicated and enforced by the Federal Elections Tribunal.

5.     Congress shall create and provide financing to support the functioning of an independent Federal Elections Tribunal to regulate the electoral process and enforce compliance with federal electoral laws and regulations.

6.     The Federal Elections Tribunal shall establish an electronic system of voter registration that will provide the means of uniquely identifying each voter, of registering each voter's date of birth and educational status, and allowing each voter to register his or her vote with strict confidentiality and security by computer, telephone, ATM or other rapid and convenient electronic means.

7.     All candidates for federal electoral office able to demonstrate the support of at least five percent of voters qualified to vote in elections to fill that office shall share equally in access to broadcast media, and in their access to financial and other forms of support that may be provided by the People through the federal government.

## THE FEDERAL DEBT

1.     Any real increase in the size of the Federal Debt of the United States over the amount of its value as of January 1, 2000 shall only be authorized by a two-thirds majority vote of both houses of Congress.

## EDUCATION AND HEALTH

1.     The United States Government shall enact and implement the necessary measures to ensure that all citizens of the United States have equal opportunity to be educated at least through four years of university or technical/vocational college, and that all citizens have

access to medical diagnostic and preventive health counseling services on an equal basis. The Government shall further ensure that all citizens have publicly-guaranteed access to curative health services and medicines up to a common, reasonable cost limit.

www.ingramcontent.com/pod-product-compliance
Lightning Source LLC
Chambersburg PA
CBHW071225290326
41931CB00037B/1970